ANIMAL SCIENCE, ISSUES AND PROFESSIONS

STUDY OF HEN'S EGGS BEHAVIOR UNDER IMPACT LOADING

Animal Science, Issues and Professions

Additional books in this series can be found on Nova's website under the Series tab.

Additional E-books in this series can be found on Nova's website under the E-book tab.

Mechanical Engineering Theory and Applications

Additional books in this series can be found on Nova's website under the Series tab.

Additional E-books in this series can be found on Nova's website under the E-book tab.

ANIMAL SCIENCE, ISSUES AND PROFESSIONS

STUDY OF HEN'S EGGS BEHAVIOR UNDER IMPACT LOADING

ŠÁRKA NEDOMOVÁ
LIBOR SEVERA
JAROSLAV BUCHAR
JAN TRNKA
AND
PAVLA STOKLASOVÁ

Nova Science Publishers, Inc.
New York

For permission to use material from this book please contact us:
Telephone 631-231-7269; Fax 631-231-8175
Web Site: http://www.novapublishers.com

LIBRARY OF CONGRESS CATALOGING-IN-PUBLICATION DATA

Hen's egg behavior under impact loading / authors, Sarka Nedomova ... [et al.].
p. cm.
Includes index.
ISBN 978-1-61761-587-0 (softcover)
1. Eggs--Experiments. 2. Eggs--Research. 3. Eggs--Packing. 4. Eggs--Production. 5. Eggshells--Experiments. 6. Eggshells--Research. I. Nedomova, Sarka, 1965-
SF490.8.H46 2010
637'.5--dc22

Published by Nova Science Publishers, Inc. † New York

CONTENTS

PREFACE

Behavior of the hen's eggs under impact loading has been investigated. Two main problems have been solved.

The first one was focused on the non – destructive impact of the egg. In this part, eggs were excited by the ball impact on the blunt side, sharp side, or the equator, and the response signals were detected by the laser-vibrometers. These sensors record the velocity of the vibration at a certain point in the direction of the laser beam. In the test, the laser beam is focused normally to the eggshell surface at a selected node on the meridian of the egg. The response wave signals were then transformed from time to frequency domain and the frequency spectrum was analyzed. The specific objectives of the research were to:

(1) analyze the response time signals and frequency signals of eggs,
(2) find the effect factors on dynamic resonance frequency, and
(3) establish relationship between the dominant frequency and the egg physical properties.

The finite element model of the egg has been developed. The eggshell is considered as linear isotropic elastic material. Its behavior is than described by the Young modulus E and by the Poisson constant ν. The numerical simulation has been performed using LS DYNA 3D finite element code. Computed signals exhibit very good agreement with experimental ones.

The second part of the research was focused on different type of impact loading when egg lying in a planar support is loaded by the falling rod. The instrumentation of the rod enables to obtain time history of the force at the point of the bar impact. The velocity of the rod is gradually increased up to certain critical value at which the eggshell failure starts. Numerical simulation of these experiments enables to obtain the stress at which the eggshell fracture occurs. This stress represents the eggshell strength. This strength is independent on the egg shape as well as on the eggshell thickness. It seems that this strength is an intrinsic material parameter which may be affected by the eggshell microstructure, by its chemical composition and by structural elements distribution. Achieved results have been used for the study of the hen's eggshell behavior at the impact on a rigid plate. Numerical results are in a reasonable agreement with records of the high speed camera.

Chapter 1

INTRODUCTION

The eggshell is the natural packing material for the egg contents, and as a result, it is important to obtain high shell strength, to resist all impacts an egg is subjected to during the production chain (Bain, 1992). Broken eggs cause economic damage in 2 ways: they cannot be sold as first-quality eggs, and the occurrence of hair cracks raises the risk for bacterial contamination of the broken egg and of other eggs when leaking, creating problems with internal and external quality and food safety. Eggshell strength is generally measured using either direct tests, such as nondestructive deformation (Voisey and Hunt, 1974) or destructive fracture force (Voisey and Hunt, 1967a, b) of an egg under quasistatic compression between 2 parallel plates, or indirect tests, such as the measurement of eggshell thickness (Brooks and Hale, 1955; Voisey and Hamilton, 1976; Ar and Rahn 1980; Thompson et al., 1981; Bell, 1984; Hunton, 1993) or specific gravity (Olsson, 1934). Many of these methods, however, are destructive, slow, or subject to environmental influences and, hence, are regarded as being unpractical. Coucke (1998) presented a fast, objective, and nondestructive method for the determination of the eggshell strength, based on acoustic resonance analysis. This research revealed that the acoustic response of the eggs to the impulse loading could be used for the identification of many physical properties of the eggs including the crack detection (Coucke, 1998; Coucke et al., 1999; De Ketelaere et al., 2000, 2002, 2003; Dunn et al., 2005 a, b). Old poultry farmers checked the mechanical integrity of hatching eggs by placing two eggs in one hand and spinning the eggs around each other while gently impacting both eggs several times. The characteristic sound produced by this impact contains information about the presence of hairline cracks in the shell. Eggs with cracked shell sound dull and highly damped and are consequently rejected from incubation. Intact eggs produce a typical sound, which is consistent when impacting on several places around the equator. In fact, by impacting the eggs in this way, they start to vibrate at one of their resonant frequencies, producing the typical sound. This test gave rise to the idea that the vibration response of eggs after impact excitation could be used to gain knowledge about the structural integrity of the eggshell. The study of the acoustic vibrations as an indicator of egg and eggshell quality is limited and relatively recent. Sinha et al. (1992) reported the use of acoustic resonant frequency analysis in chicken eggs for the detection of *Salmonella enteriditis* bacteria in the egg content. The excitation was done at the blunt side of the egg with a piezoelectric crystal and the response is measured with an accelerometer at the opposite side. Yang et al. (1995) used the vibrational behavior of a chemically treated egg as a quality detection method. Bliss (1973) and Moayeri (1997) detect the local stiffness of the eggshell by analyzing the time signal of the impactor after impacting. Multiple measurements distributed over the whole eggshell surface are required to have a global overview of the eggshell strength.

The present study was intended to better understanding the vibration characteristics of an egg. For this purpose, an experimental modal analysis was performed on an intact egg. This analysis gives the possibility to visualize the spatial motion of the egg after being impacted. In order to achieve these goals, eggs were excited by the impact of the steel ball on the blunt side, and the response signals were detected by the laser vibrometers at the different points on the eggshell surface. The response wave signals were then transformed from time to frequency domain and the frequency spectrum was analyzed. The specific objectives of the research were to:

- analyze the response time signals and frequency signals of eggs
- develop a finite element model of an egg.

The next part of the presented paper deals with the problem of the eggshell strength evaluation under impact loading. A new experimental technique has been developed. Numerical simulation of this method has also been performed. This procedure enables to determine the eggshell strength in terms of the stress. Experiments focused on the hen's eggshell behavior at the impact on a rigid plate have been used for the verification of the obtained model of an egg. By this way the complex behavior of the eggshell at impact loading has been described.

Chapter 2

Eggshell Behavior at Non - Destructive Impact

2.1. Egg Samples

Eggs were collected from a commercial packing station. The physical properties of egg samples were determined by the following methods: linear dimensions, i.e. length (L) and width (W), were measured with a digital calliper to the nearest 0.01 mm. The geometric mean diameter of eggs was calculated using the following equation given by (Mohsenin, 1970):

$$D_g = \left(LW^2\right)^{\frac{1}{3}} \tag{1}$$

According to Mohsenin (1970), the degree of sphericity of eggs can be expressed as follows:

$$\Phi = \frac{D_g}{L} x100 \quad (\%) \tag{2}$$

The surface area of eggs was calculated using the following relationship given by (Mohsenin, 1970; Baryeh and Mangope, 2003):

$$S = \pi D_g^2 \tag{3}$$

Volume of the egg is than given as

$$V = \frac{\pi}{6} LW \tag{4}$$

Shape index of the eggs was determined using following equation:

$$SI = \frac{W}{L} x100 \quad (\%) \tag{5}$$

The main properties of the eggs are presented in the Table 1.

Table 1. Main characteristics of the tested eggs

No.	Mass (g)	Width (cm)	Length (cm)	Shape index (%)	Air cell (mm)
1	60.39	44.71	54.16	82.55	2
2	60.75	45.28	53.50	84.64	1
3	66.50	43.99	61.10	72.00	2
4	65.17	45.21	56.32	80.27	1
5	61.73	44.74	55.50	80.61	1
6	64.17	44.60	57.33	77.80	2
7	65.79	45.20	57.21	79.01	2
8	63.95	44.21	57.36	77.07	2
9	68.66	46.03	57.18	80.50	2
10	64.30	45.99	54.03	85.12	2
11	67.51	45.69	57.52	79.43	1
12	66.67	45.45	57.63	78.87	2
13	65.30	45.02	57.70	78.02	1
14	60.39	43.51	58.23	74.72	2
15	63.86	44.33	58.73	75.48	2
16	63.25	44.62	55.93	79.78	2
17	60.84	43.91	56.26	78.05	2
18	62.17	45.06	54.86	82.14	2
19	62.54	43.78	58.51	74.82	2
20	63.13	44.20	58.53	75.52	2
21	62.92	44.51	56.37	78.96	2
22	64.26	45.62	54.72	83.37	2
23	65.54	44.90	58.03	77.37	2
24	67.79	46.20	56.89	81.21	2
25	61.48	44.29	55.28	80.12	1
26	66.77	45.56	56.93	80.03	2
27	63.43	44.43	56.48	78.67	2
28	68.54	45.81	57.91	79.11	2
29	64.37	44.41	58.05	76.50	2
30	66.44	45.19	57.94	77.99	2
average	64.29	44.88	56.87	78.99	1.80

Although the shape index given in the Table 1 may be sufficient for the description of many problems – see e.g. (Altuntas and Seroglu, 2008), the solution of stress and strain state in the eggshell must be based on the exact description of the eggshell shape. There are many procedures which can be used for the solution of this task. One of the most popular is the mathematical equation given by Narushin (2001) for the profile of an egg. This equation (Eq. 6) adequately defined the surface profile of the egg based on the length (L) and the maximum breadth (B) of the egg:

$$y = \pm\sqrt{L^{\frac{2}{n+1}} x^{\frac{2n}{n+1}} - x^2} \tag{6}$$

where:

$$n = 1.057\left(\frac{L}{B}\right)^{2.372} \tag{7}$$

There are many other equations – see (Erdogdu et al., 2005) for a review. More accurate determining of egg's shape is based on the analysis of the digital egg photographs. The application required one measured dimension (e.g. the egg length, measured with sliding calliper), and allowed the user to determine any user defined distance on the photograph from the derived number of pixels per unit length. From the dimensional measures of individual eggs, their contours could be accurately described in a user defined Cartesian coordinate system, using a mathematical equation. For more details, the reader is referred to the procedure described by Denys et al. (2003). Three dimensional egg shapes can be then obtained by revolving the contours 180° about the axis of symmetry. The shape of the eggshell counter can be described using of the polar coordinates r,φ as:

$$x = r\cos\varphi \qquad y = \sin\varphi, \tag{8}$$

where

$$r(\varphi) = a_o + \sum_{i=0}^{\infty} a_i \cos\left(2\pi\frac{\varphi}{c_i}\right) + b_i \sin\left(2\pi\frac{\varphi}{c_i}\right) \tag{9}$$

The analysis of our data led to the conclusion that the first four or five coefficients of the Fourier series are quite sufficient for the egg's counter shape description (the correlation coefficient between measured and computed egg's profiles lies between 0.98 and 1).

In the Figure 1a an example of the egg counters computed using of the Eq. (6) and determined from the digital photography is displayed. It can be seen that there is a difference between these two counters. The knowledge of the equation describing the eggshell counter is necessary namely for the numerical simulation of egg behavior under different mechanical loading, at numerical simulation of different heat treatment and also for the determination of the curvature of this curve. The radius of the curvature, R, than plays meaning role at the evaluation of some egg's loading tests (compression test, etc.) – see e.g. MacLeod et al. (2006). This radius can be solved by the solving of the following system of equations:

$$
\begin{aligned}
&F(\varphi) = (x - x_o)^2 + (y - y_o)^2 - R^2 \\
&F(\varphi) = 0 \\
&\frac{dF(\varphi)}{d\varphi} = 0 \\
&\frac{d^2 F(\varphi)}{d\varphi^2} = 0
\end{aligned}
\tag{10}
$$

where x_o, y_o are the coordinates of the center of the curvature of the given curve. If the curve is described as function y=f(x) – see Eq. 6, than the radius of the curvature is given as:

$$
R = \frac{\left(1 + \left(\frac{dy}{dx}\right)^2\right)^{\frac{3}{2}}}{\frac{d^2 y}{dx^2}}
\tag{11}
$$

An example of the radius of the eggshell contour curvature is displayed in the Figure 2.

Examples of the parameters $a_0 - a_5$ are given in the Table 2.

Table 2. Parameters of the Fourier series

Egg No.	a_0 (mm)	a_1 (mm)	a_2 (mm)	a_3 (mm)	a_4 (mm)	a_5 (mm)
1	2.12499	-0.29335	0.34217	-0.12421	0.17387	0.09230
2	2.09077	-0.48952	0.34469	-0.17596	0.32507	0.00512
3	2.20550	-0.22044	0.35509	0.00817	0.24207	0.02732
4	2.07706	-0.53085	0.46642	-0.26738	0.17613	0.13600
5	2.10414	-0.37483	0.52265	-0.16073	0.16685	0.08387
6	2.21139	0.08756	0.41620	-0.18620	0.23424	0.33348
7	2.11313	-0.30195	0.39717	0.01323	0.18090	0.08909
8	2.25341	-0.32390	0.36839	-0.00570	0.22664	0.08596
9	2.09777	-0.56694	0.44252	-0.20463	0.24923	0.04899
10	2.10381	-0.26255	0.35309	-0.02600	0.30705	-0.00028
11	2.14153	-0.42771	0.46006	-0.24329	0.31052	0.16553
12	2.08970	-0.47290	0.46277	0.15292	0.17199	-0.16723
13	2.25533	0.43566	0.05081	0.54100	0.58856	-0.22585
14	1.98390	-0.60629	0.54410	-0.36056	0.21227	0.08132
15	2.04938	-0.30450	0.38307	-0.09057	0.15791	0.07845
16	2.02310	-0.56432	0.56001	-0.18340	0.24989	-0.01039
17	1.90091	-0.60728	0.66132	-0.47316	0.27122	0.11678
18	2.00963	-0.64351	0.39148	-0.23410	0.30908	-0.01978
19	2.26103	-0.20451	0.27660	0.01392	0.17835	-0.03563
20	2.05119	0.51333	0.64971	-0.06500	0.05849	0.22746
21	2.02030	-0.39033	0.55991	-0.11977	0.31929	0.05467
22	2.11194	-0.22980	0.38933	0.09574	0.13560	0.02557
23	2.19283	-0.23854	0.36055	-0.11654	0.18820	0.07914
24	2.17823	-0.18673	0.30867	0.01258	0.07538	0.02141
25	2.15933	-0.13679	0.38355	-0.06446	0.15137	0.13713
26	2.16613	-0.27726	0.32321	-0.34317	0.24034	0.16623
27	2.00980	-0.53561	0.55880	-0.47912	0.20549	0.22067
28	2.16542	-0.15770	0.32021	0.15656	0.25722	0.05457
29	2.07503	-0.25397	0.49153	-0.15944	0.16201	0.08691
30	2.15128	0.00970	0.30886	-0.07292	0.19626	0.13395

Figure 1 shows an example of the egg's shape obtained using Eqs. (6) and (8). A significant difference can be seen. It means that the exact evaluation of the egg's shape is needed.

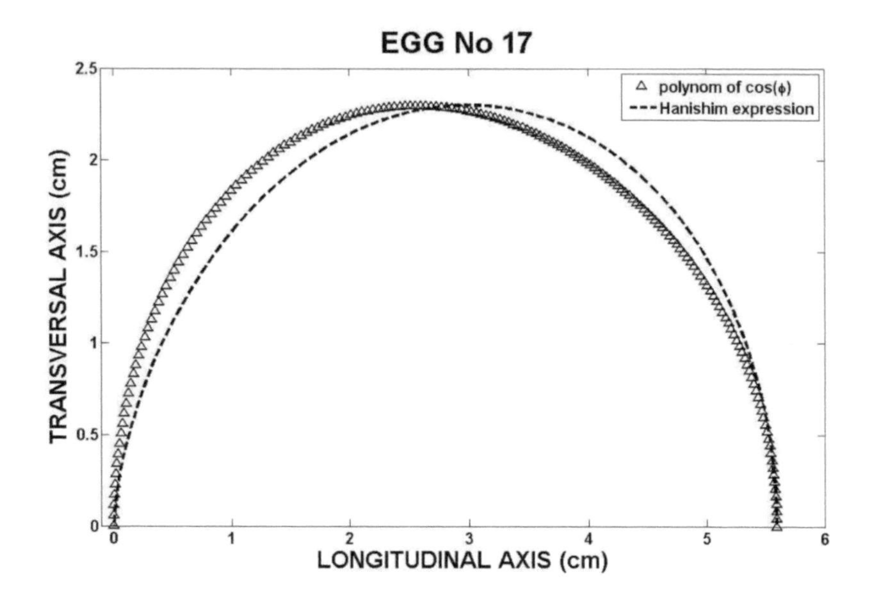

Figure 1. Egg's profile determined using Equations (6) and (8).

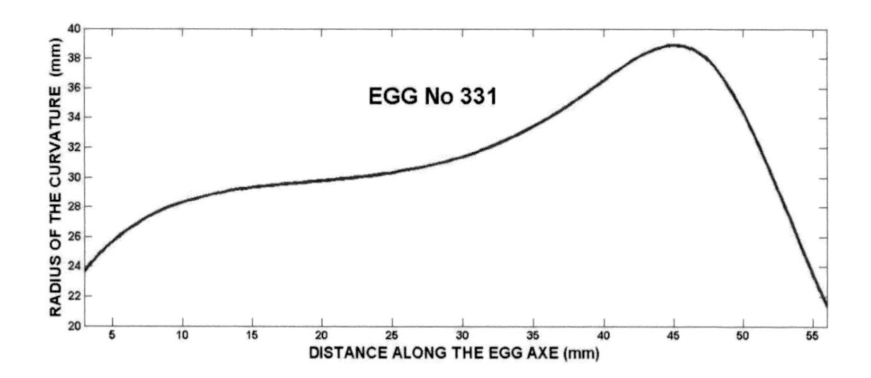

Figure 2. Variation of the radius of the curvature along the egg's profile.

2.2. Experimental Method

The measurement set-up is shown in Figure 3. The system has three major parts, namely the product support, the excitation device and the response-measuring device.

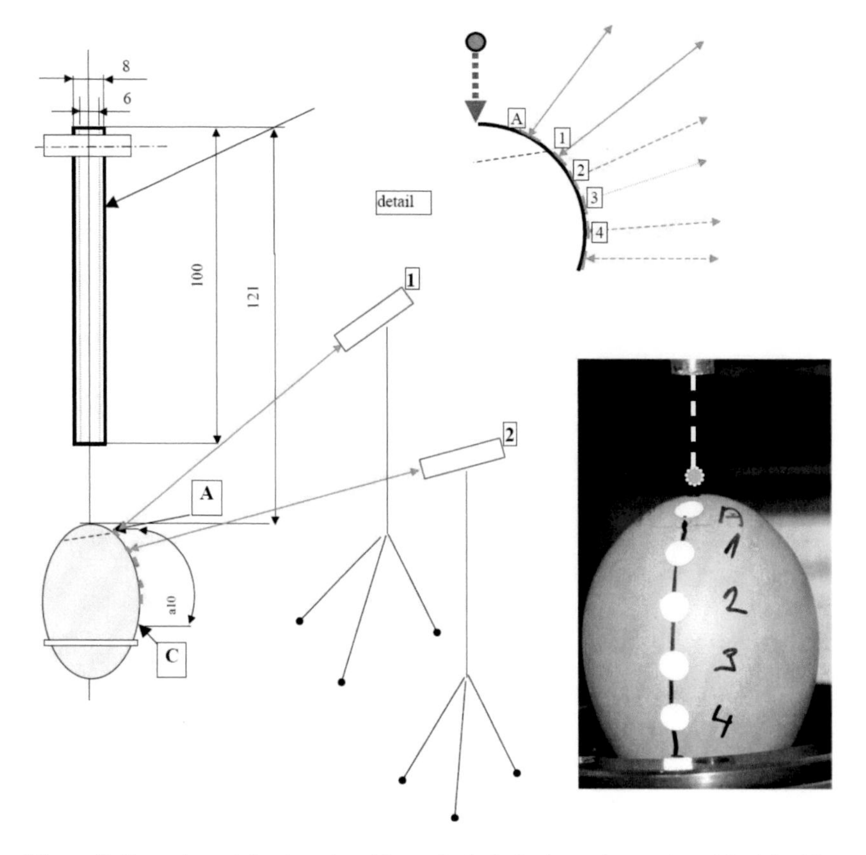

Figure 3. Experimental set up (positions A, 1, 2, 3, 4 are denoted as A0, A1, A2, A3, A4).

Product support - During the measurements the support of the egg must be such that the distortion of its natural motion caused by this support is minimal. A teflon ring was chosen as a supporting mean.

The impact excitation method was chosen because of its fast and simple nature. The egg is excited at top of the blunt part of the egg by the impact of the steel ball. The ball (6 mm in diameter) falls from the height of 121 mm.

The egg response measurement. A laser vibrometer was used to measure the egg response to the impact. This contactless sensor adds no extra mass to the structure and does not disturb the free vibration of the egg. The laser-vibrometer measures the velocity of the vibration at a certain point in the direction of the laser beam. In the test, the laser beam is focused normally to the eggshell surface at a selected node on the meridian of the egg. The laser vibrometer is isolated from the egg supporting structure so no disturbing vibrations are introduced when performing the measurements.

Data acquisition and analysis. When eggs were excited, the response acceleration signals in time domain were detected, and MATLAB computer program was used to transform the response from time to frequency domain, by means of FFT. Dynamic response curves were then gained and statistically analyzed in the time and frequency domain for all eggs. The experiments were conducted with five replicates.

2.3. Experimental Results

2.3.1. Time Domain

In the Figure 4 an example of the surface velocity of an egg surface is given. It can be seen that the reproducibility of the experiments is relatively very high. Pronounced damping of the time signal is clearly visible. The damping is also shown in the Figure 5. The positions of the single points along the meridian are given in the Table 3.

Table 3. Position of the detection points along the meridian (Egg No. 19)

Point	Distance from the blunt pole (mm)	ANGLE α (°)
A	7.4	14.45
1	18.7	37.63
2	28.7	59.74
3	38.8	18.01
4	48.4	109.46

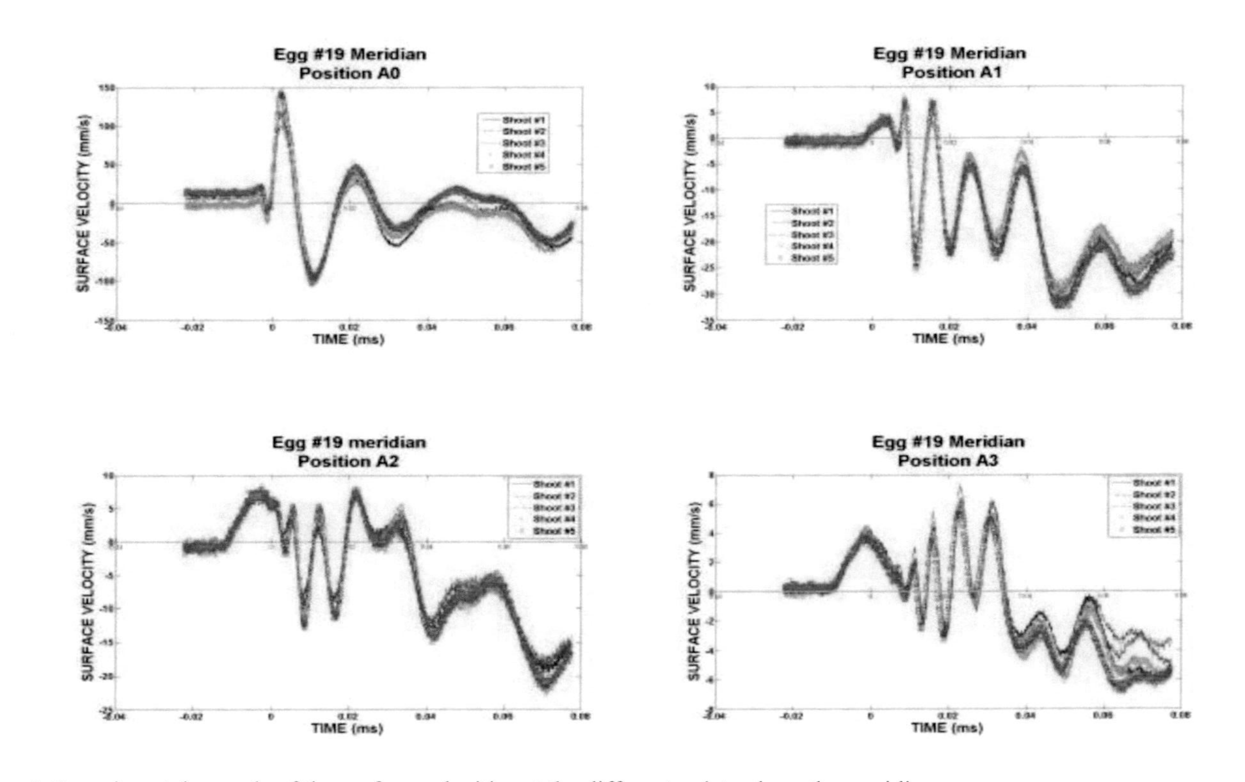

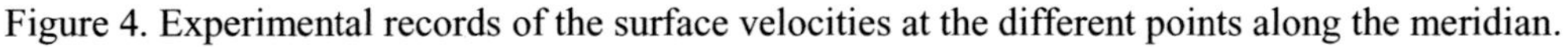
Figure 4. Experimental records of the surface velocities at the different points along the meridian.

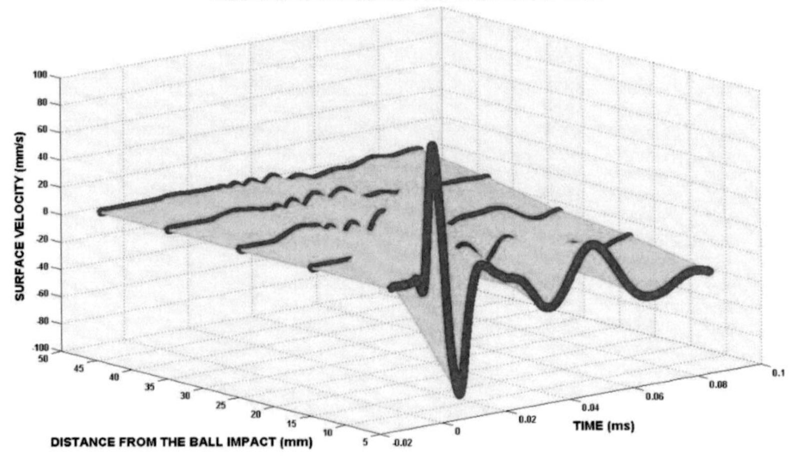

Figure 5. Development of the surface velocity along the meridian.

The damping of the surface velocity can be also expressed as the dependence of the surface velocity maximum on the distance from the blunt pole – see Figure 6.

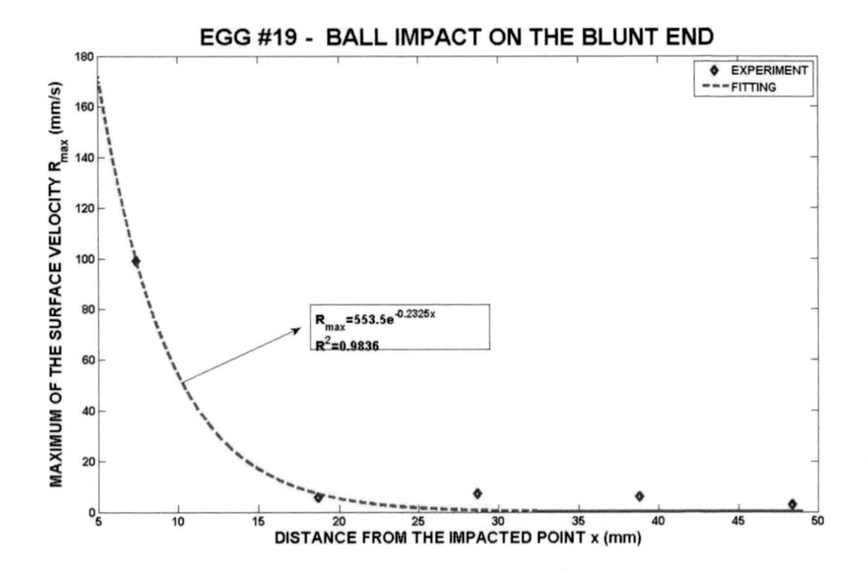

Figure 6. Damping of the surface velocity along the meridian.

The surface velocity corresponds to the surface wave which is initiated at the blunt pole by the ball impact. The velocity of this wave can be evaluated. The dependence of this velocity on the distance from the point of the ball impact is shown in the Figure 7. The decrease in this velocity is a consequence of the egg surface curvature, contact with the egg membranes, egg liquids etc. The investigation of these effects must be based on further experiments.

The remaining eggs have been tested by the same way as the egg No. 19. The response has been recorded at the position A0 and on the equator. Examples of these records are shown in Figure 8. The recorded signals are very similar. The surface wave velocities have also been evaluated. Results are given in Table 4.

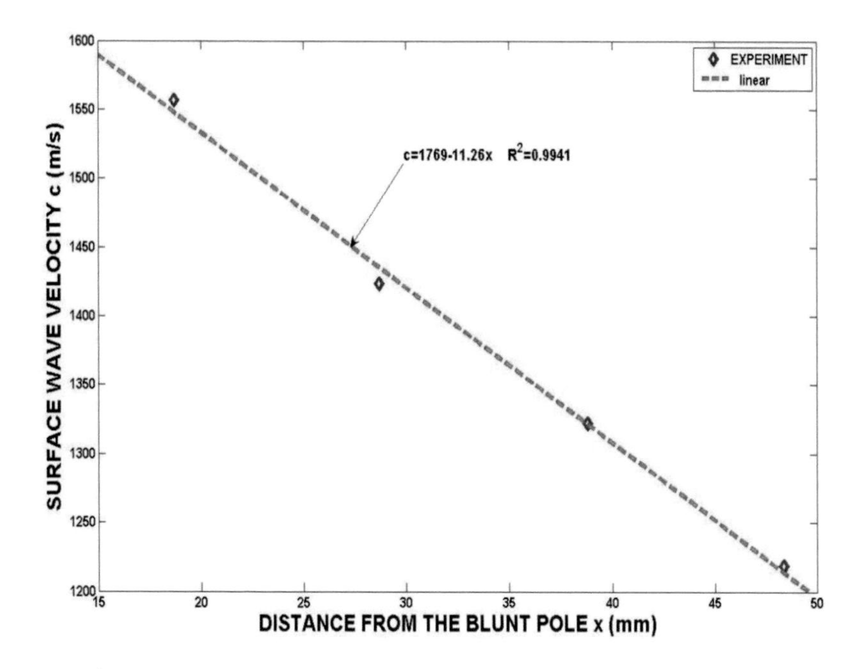

Figure 7. The decrease of the surface wave velocity with the distance from the blunt pole (Egg No. 19).

Table 4. Surface wave velocities

Egg No.	c (m/s)	Egg No.	c (m/s)	Egg No.	c (m/s)
1	1219.3	11	1226.3	21	1221,6
2	1222.5	12	1221.6	22	1218.3
3	1223.8	13	1217.6	23	1219.1
4	1218.6	14	1224.3	24	1222.7
5	1217.3	15	1227.1	25	1225.7
6	1219.1	16	1215,8	26	1218.4
7	1222.4	17	1216,3	27	1219.5
8	1220.4	18	1214,2	28	1221.7
9	12222.6	19	1218,2	29	1223.4
10	1219.3	20	1225.4	30	1224.1

The distribution of these velocities is displaced in the Figure 8.

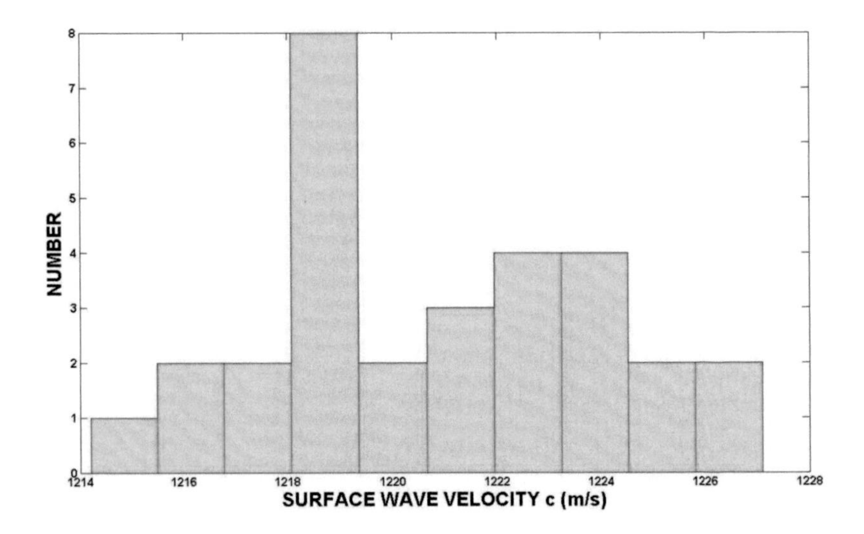

Figure 8. Distribution of the surface wave velocities.

2.3.2. Frequency Domain

The transition from the time to the frequency domain has been performed using the fast Fourier transform (FFT). This transform enables to express a time function f(t) as:

$$f(t)=\frac{1}{\pi}\int_{=\infty}^{\infty}S(\omega)e^{i\omega t}d\omega \qquad S(\omega)=\int_{-\infty}^{\infty}f(t)e^{i\omega t}dt \quad , \quad (12)$$

where S(ω) is the spectral function and ω denotes the angular frequency. The transform into the frequency domain will be a complex valued function, that is, with magnitude and phase. The fast Fourier technique (FFT) has been used for the evaluation of the magnitude and phase. This algorithm is a part of the MATLAB software. Examples of the amplitudes of the spectral functions of the surface velocities are shown in the Figs. 9-11. The amplitudes are significant namely for a lower frequencies. It is evident that the maximum of the spectral function amplitude decreases with the transition from the point A0 (near of the blunt end) to the equator. This decrease can be expressed using of the transfer function T(ω) which is defined as ratio:

$$T(\omega)=\frac{S(\omega)|x=EQUATOR}{S(\omega)|x=A0} \quad (13)$$

Examples of the amplitudes of the transfer functions are displaced in the Figure 12.

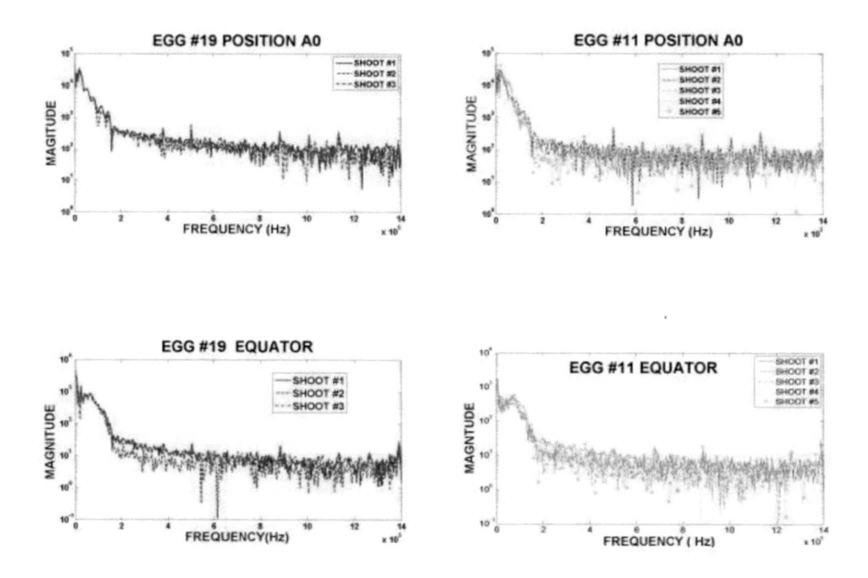

Figure 9. Amplitudes of the spectral functions.

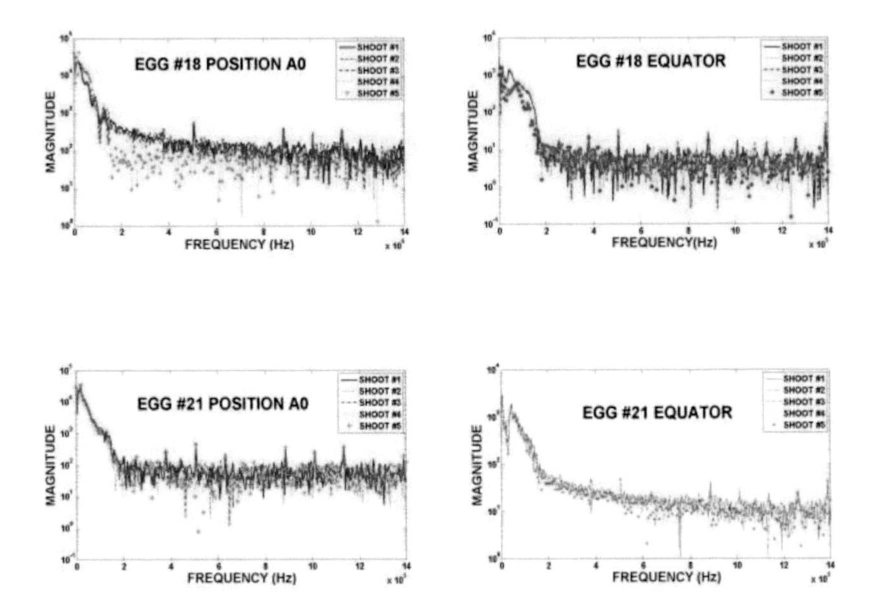

Figure 10. Amplitudes of the spectral functions.

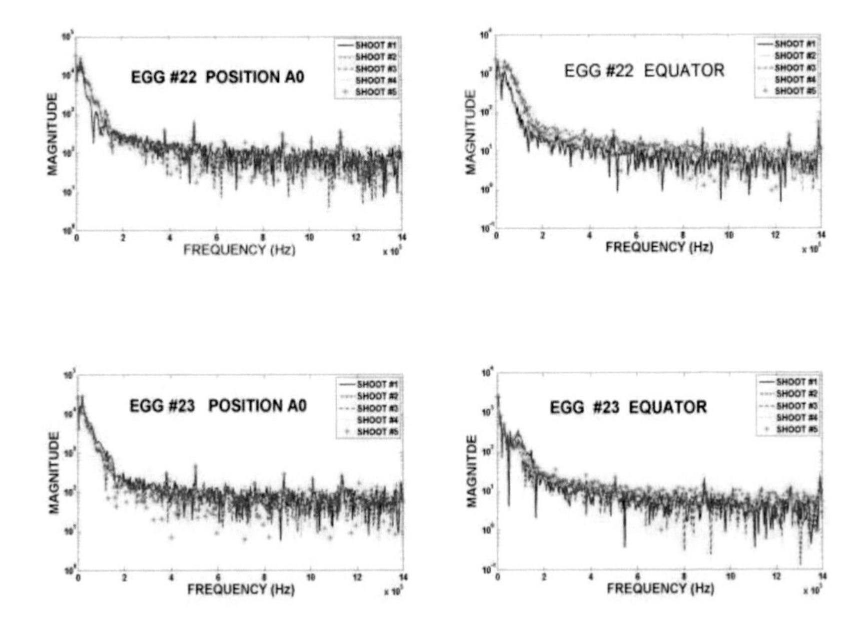

Figure 11. Amplitudes of the spectral functions.

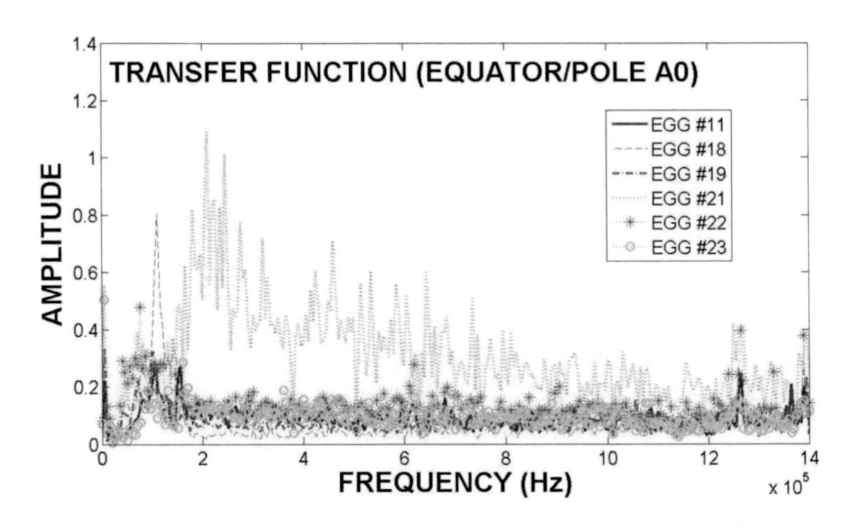

Figure 12. Examples of the transfer functions.

More detailed view on the changes of the spectral functions in the direction of the surface wave propagation can be seen in the Figure 13.

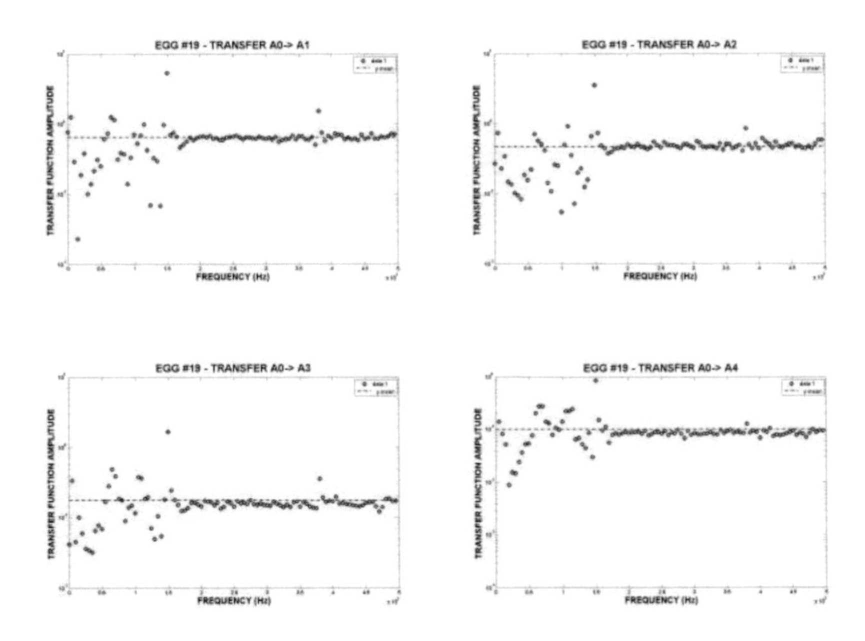

Figure 13. Transfer function amplitude around the meridian – egg No. 19.

General view on the transfer functions along the meridian is displaced in the Figure 14. There are certain frequencies at which some amplification of the spectral functions can be observed. In these figures the average values of the transfer functions amplitude are also shown. The average amplitude is decreasing function of the distance from the blunt pole – see Figure 15.

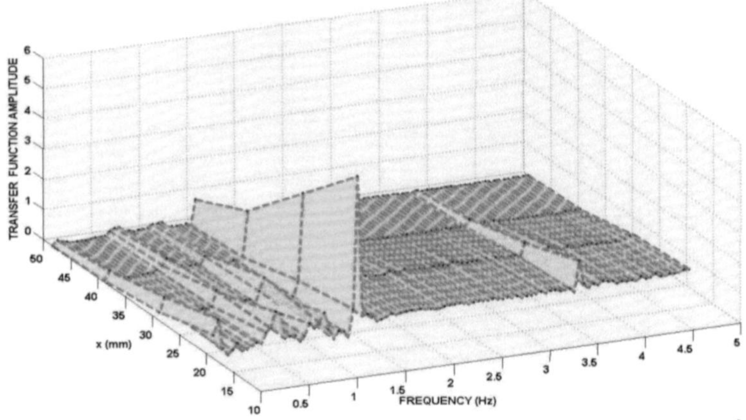

Figure 14. Transfer function amplitudes.

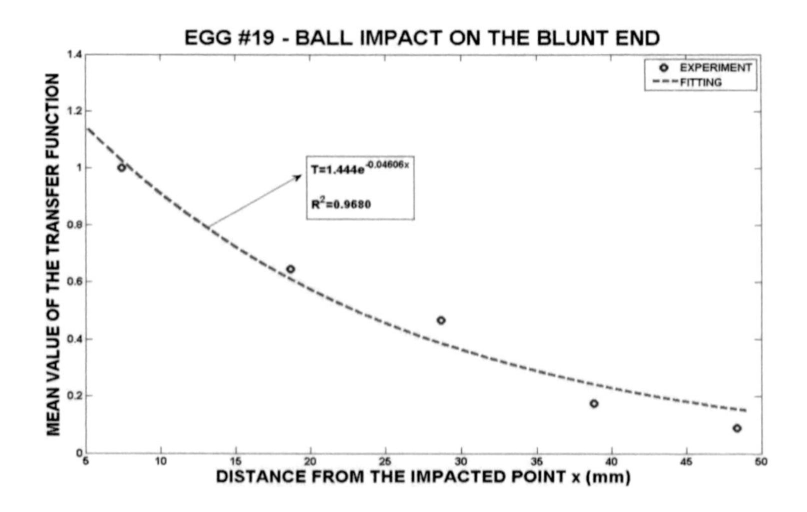

Figure 15. Average values of the transfer functions amplitudes.

The average values for all tested eggs are given in the Table 5.

Table 5. Average values of the transfer functions (blunt pole - equator) T

Egg No.	Shape Index (%)	T
1	82.55	0.16532
2	84.64	0.18112
3	72.00	0.04892
4	80.27	0.09021
5	80.61	0.09032
6	77.80	0.06451
7	79.01	0.08695
8	77.07	0.06594
9	80.50	0.08976
10	85.12	0.18640
11	79.43	0.07983
12	78.87	0.07234
13	78.02	0.07562
14	74.72	0.05631
15	75.48	0.08204
16	79.78	0.08917
17	78.05	0.08192
18	82.14	0.16000
19	74.82	0.10850
20	75.52	0.07146
21	78.96	0.08361
22	83.37	0.17231
23	77.37	0.05983
24	81.21	0.09235
25	80.12	0.08967
26	80.03	0.09023
27	78.67	0.08236
28	79.11	0.08712
29	76.50	0.07153
30	77.99	0.06611

The average value of the transfer function amplitude increases with the increase of the shape index. This tendency is shown in the Figure 16. It means that approach to the spherical shape leads to improvement of the transfer capability of the eggshell.

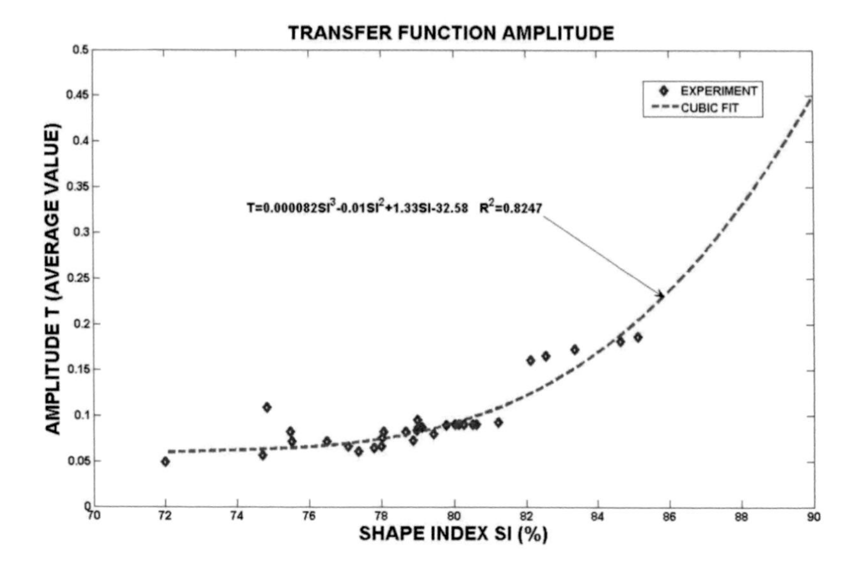

Figure 16. The dependence of the transfer function amplitude on the egg's shape.

Obtained results show that the best position for the egg excitation is the blunt pole. The response function of the egg should be detected namely in the area of the air cell.

2.4. Numerical Simulation

The numerical model should reflect the main features of the egg structure. Schematic of the hen's egg is shown e.g. in Sugino et al. (1997). The detail description of the single elements of this structure is given e.g. in Wells (1968), Rodrigez-Navarro et al. (2002). The finite element model has been developed using the following assumption:

Eggshell is homogeneous isotropic linear elastic material. The properties of such material are described by the Young modulus E, Poisson ratio ν and material density ρ.

Membranes are also taken as linear elastic material. No difference between membranes has been considered.

Air is considered as an ideal gas.

Egg yolk and egg white are considered as compressible liquids.

Cross section of the used model is shown in the Figure 17. The numerical model is shown in the Figure 18.

Parameters of the model are:

- Total number of nodes 141249
- Total number of solid elements 88050
- Total number of shell elements 46710

Numerical analysis has been performed by LS DYNA 3D finite element code.

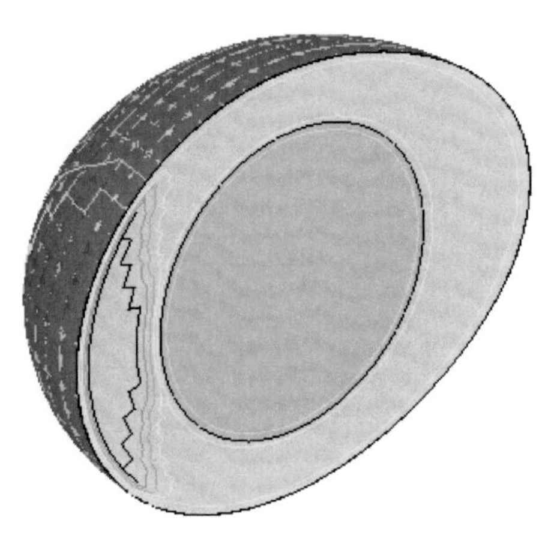

Figure 17. Schematic of the egg model used for the numerical simulation.

The elastic properties of the eggshell have been obtained using of the procedure developed in Buchar and Simeonovová (2001). Elastic properties of the membranes were determined by the method described in Bing Feng Ju et al. (2002). The compressibility of the egg liquids was taken from the study Chung and Stadelman (1965). The elastic properties are given in Table 6 and 7.

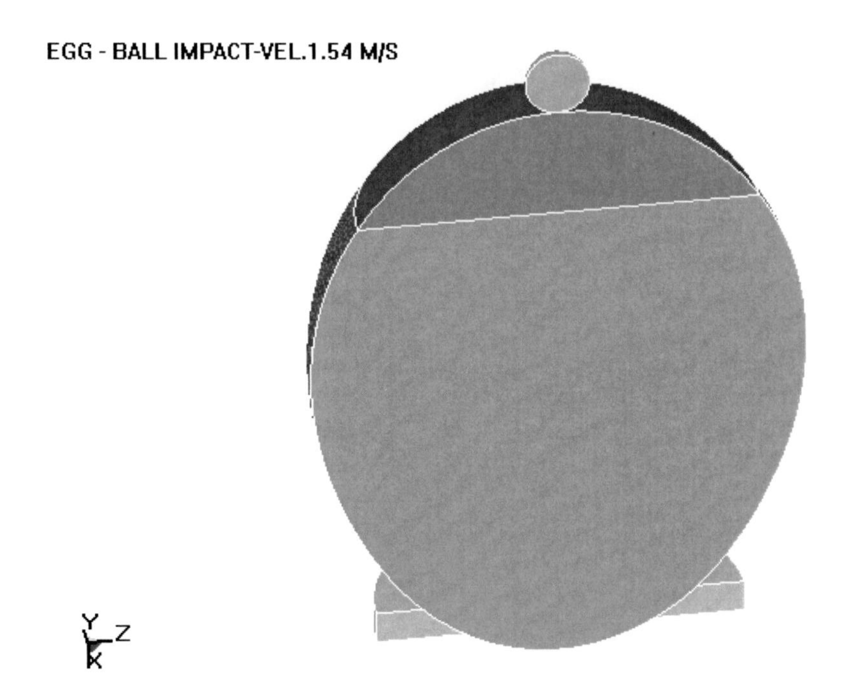

Figure 18. Numerical model of the experiment.

Table 6. Elastic properties of the egg parts

Egg part	ρ (kg/m^3)	E (GPa)	ν
eggshell	2140	20.8	0.37
membrane	1005	0.0035	0.45

Table 7. The properties of the egg liquids. (K- bulk modulus)

Egg liquid	ρ (kg/m^3)	K (GPa)
white	2140	2.0
yolk	1005	1.8

The surface velocities have been evaluated at different nodes corresponding to the points where the experimental data have been reported. A schematic of these nodes is shown in the Figure 19.

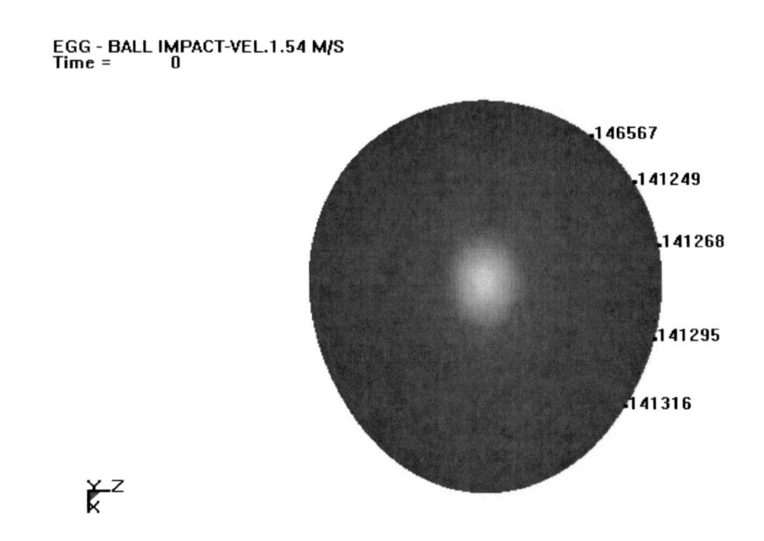

Figure 19. Nodes where the velocities have been computed. Single nodes correspond to the point A0, A1, A2, A3, and A4.

The surface velocity is computed in the x and y directions. The velocity normal to the eggshell surface has been evaluated by the procedure outlined in Figure 20a-e. The five versions have been considered. Examples of the numerical computations are displayed in the following figures.

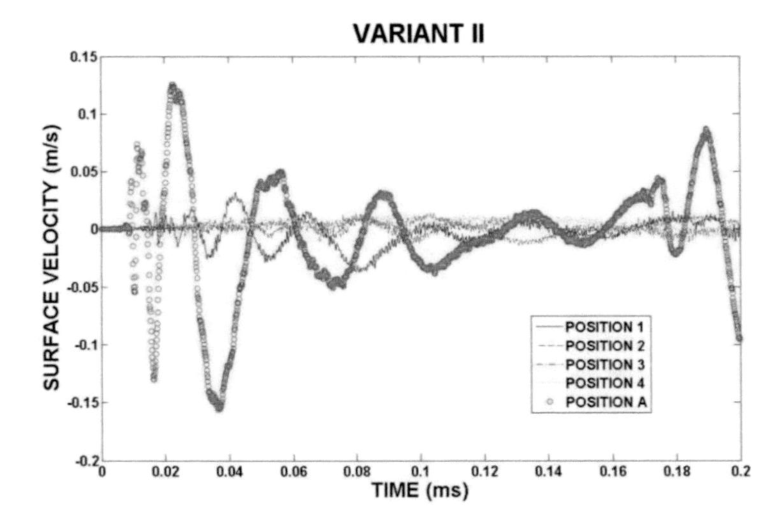

Figure 20a. Schematic of the surface velocity evaluation. Variant I. Egg is freely supported by the teflon ring.

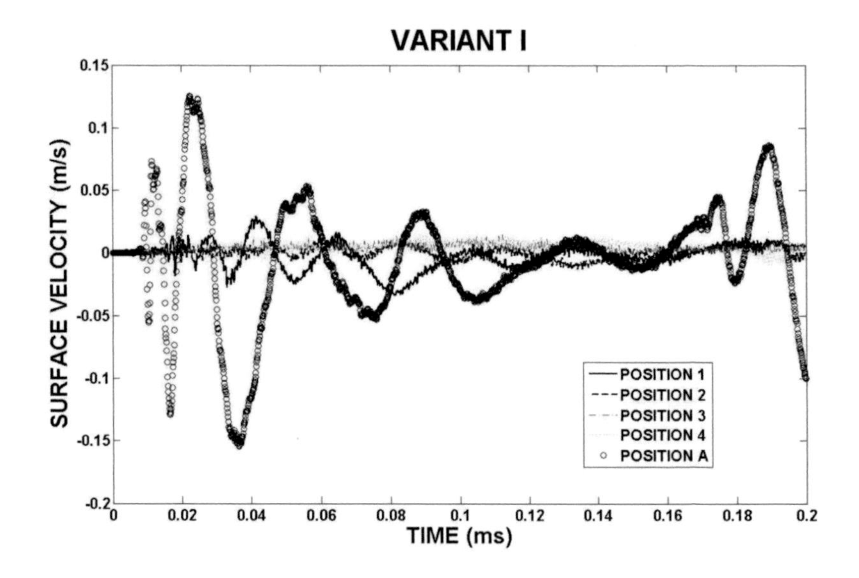

Figure 20b. Schematic of the surface velocity evaluation. Variant II. Egg is connected with the teflon ring.

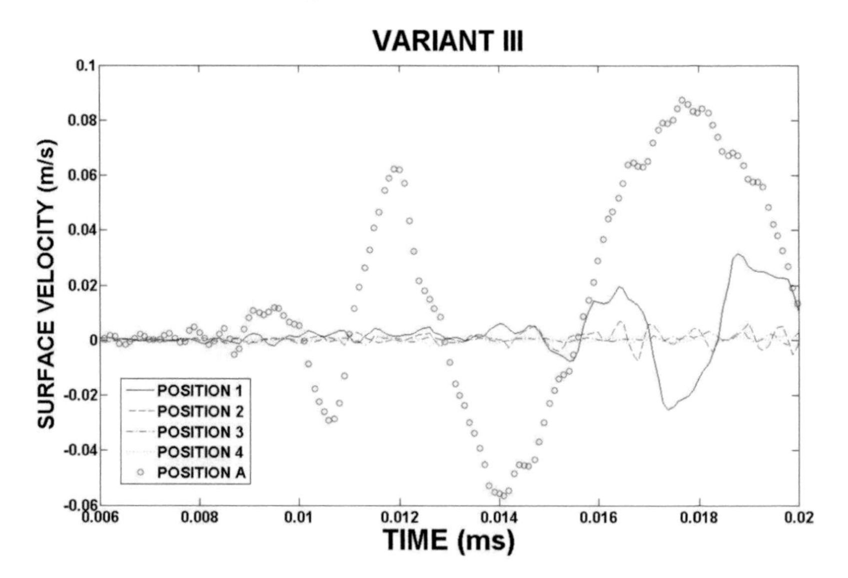

Figure 20c. Schematic of the surface velocity evaluation. Variant III. Egg is in the contact with a rigid plate.

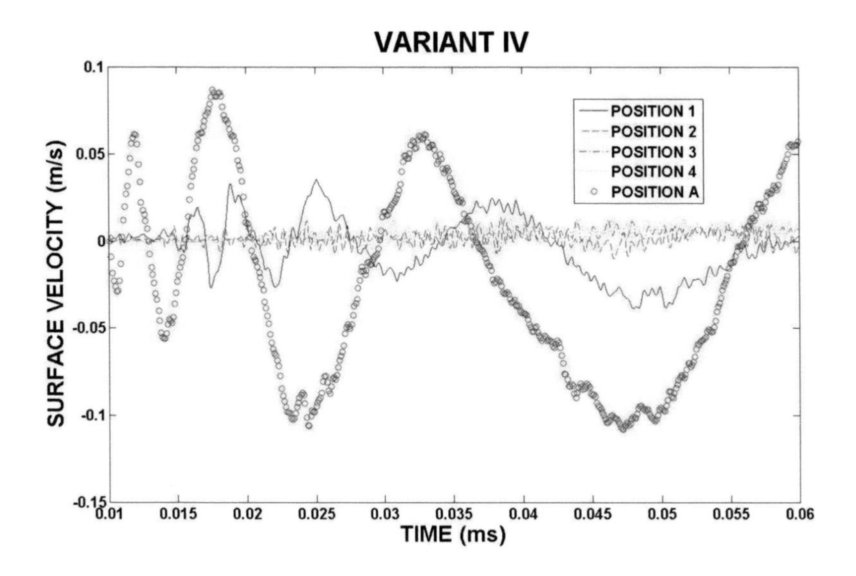

Figure 20d. Schematic of the surface velocity evaluation. Variant IV. No ring is considered.

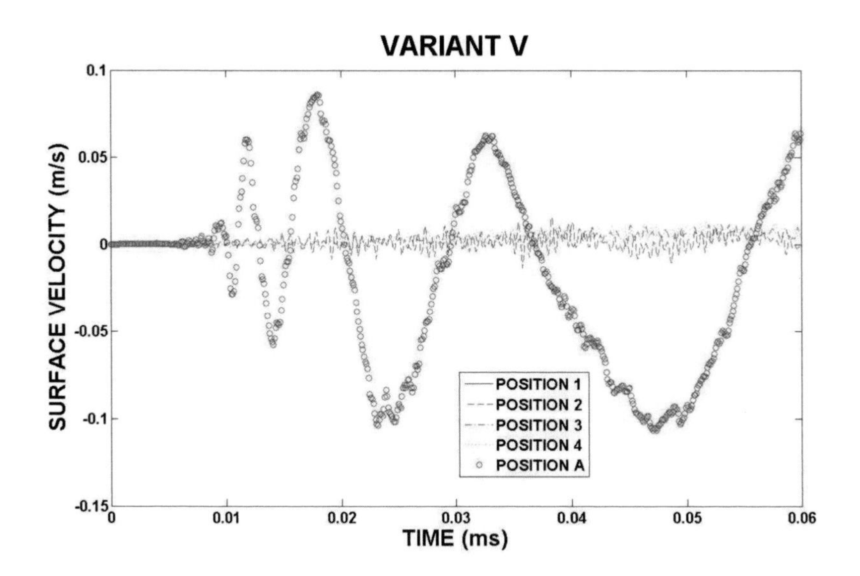

Figure 20e. Schematic of the surface velocity evaluation. Variant V. Egg is only supported at the sharp end.

It seems that the best agreement with the experimental results was obtaine in the case of variant I. It is evident that the boundary value conditions play meaningful role. The process of the impact loading cannot be considered as a local event. It means it is necessary to use more exact description of the egg liquids behavior. The reasonable agreement between numerical and experimental response functions approves using of the model developed in the given paper for solution of impact problems.

The results obtained during experiments on the non-destructive impact show that the acoustic methods can be considered as very effective tool for the study of the egg's behavior under different loading conditions. In the next step the new method of the dynamic strength of the eggshell has been developed.

Chapter 3

EGGSHELL STRENGTH UNDER IMPACT LOADING

To study the egg resistance to impact, many methods have been developed - see e.g. Tyler and Geake (1963), Voisey and Hunt (1967, 1968). These methods use ball or rod, which are dropped on the eggshell and the height of fall, the size of ball or a number of blows are used to estimate shell strength. Within the work presented in this chapter, the new experimental method of the dynamic strength evaluation has been developed. Loading has been performed by the impact of the free falling rod. The record of the force at the point of rod-eggshell contact enables to evaluate the rupture force at a definite impact velocity. Obtained results have been compared with results of the static compression of the eggs in order to find possible evidence of the loading rate influence.

The numerical simulation of the impact experiments have been used for evaluation of stresses at the moment of the egg's break.

3.1. EXPERIMENTAL DETAILS

Eggs (*Hisex Brown* strain) were collected from a commercial packing station. Typically, eggs were maximal 2 days old when they arrived at the grading station. The main characteristics of the eggs have been evaluated. These characteristics are given in the Table 1.

The experimental set-up is very similar to that shown in Figure 3 – see Figure 21. It also consists of three major components; they are the egg support, the loading device and the response-measuring device.

- The egg support is a cube made of soft polyurethane foam. The stiffness of this foam is significantly lower than the eggshell stiffness; therefore there is very little influence of this foam on the dynamic behavior of the egg.
- A bar of the circular cross-section with strain gauges (semi conducting, 3 mm in length) is used as a loading device. The bar is made from aluminum alloy. Its length is 200 mm, diameter is 6 mm. The bar is allowed to fall freely from a pre-selected height. The instrumentation of the bar by the strain gauges enables to record time history of the force at the area of bar-eggshell contact.
- The response of the egg to the impact loading, described above, has been measured using the laser vibrometer. This device enables to obtain the time history of the eggshell surface displacement.

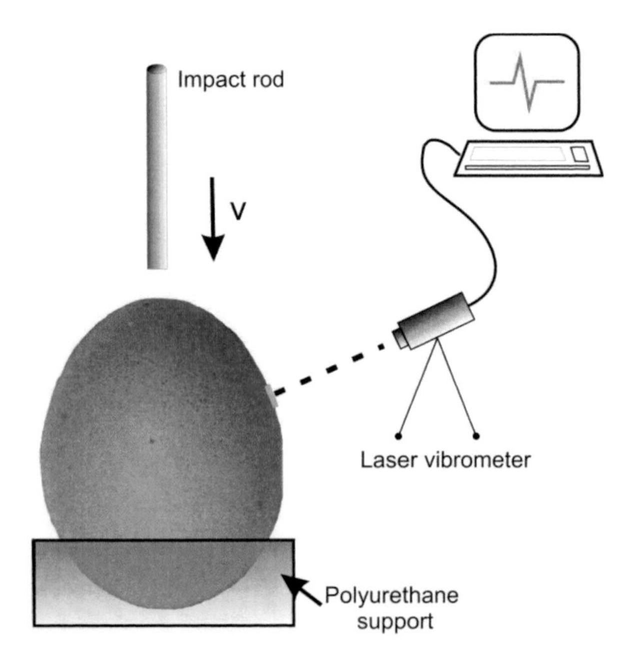

Figure 21. Schematic of the impact loading of the egg.

The eggs have been impacted on the sharp end, on the blunt end, and on the equator. The height of the bar fall has been increased up to value at which the eggshell damage has been observed. The displacement has been recorded on the equator of the egg. The displacement has been measured in normal direction to the eggshell surface.

The static compression has been performed using the test device (TIRATEST 27025, Germany) which has three main components: a stationary and a moving platform, and a data acquisition system. Compression force was measured by the data acquisition system. The egg was placed at the block of polyurethane foam positioned on the stationary plate - as it is shown in the Figure 21. The egg has been loaded by the moving rod (6 mm in diameter) at a speed of 20 mm/min. Two compression axes (*X* and *Z*) for the egg were used in order to determine the rupture force and deformation. The *X*-axis was the loading axis through the length dimension, and the *Z*-axis was the transverse axis containing the width dimension. Along the *X* - axis two other orientations have been considered. The eggs have been loaded at the sharp end and at the blunt end. For each orientation 30 eggs have been tested.

3.2. Experimental Results

3.2.1. Static Compression

In Figure 22 the examples of experimental records of the force vs. rod displacements for the different loading axes are shown.

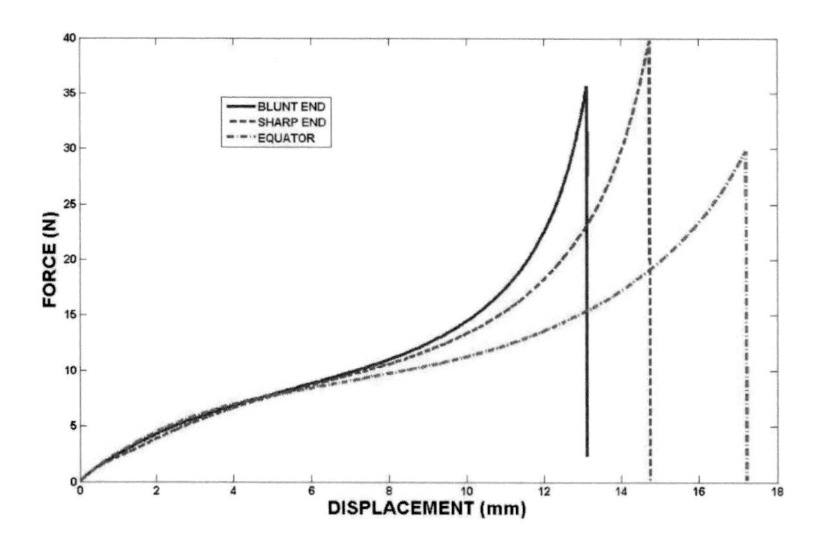

Figure 22. Experimental records of the force – displacements for the different loading orientation.

It can be seen that the shape of these curves is different from those obtained at the eggs compression between two plates - see Altuntas and Sekeroglu (2008). The observed dependences can be fitted by the polynomial:

$$F = a_1x^4 + a_2x^3 + a_3x^2 + a_4x + a_5 \tag{14}$$

where x is the rod displacement in millimeters. For all the experiments the correlation between this fit and experimental record is better than 0.98.

The egg stiffness at this type of loading can be expressed using the *dF/dx* function. Its course is shown in Figure 23. One can see that this stiffness exhibits more or less significant dependence on the rod displacement. The stiffness at the egg compression between two plates is nearly constant. The maximum of the force has been taken as the rupture force. This force can be considered as the strength of the eggshell.

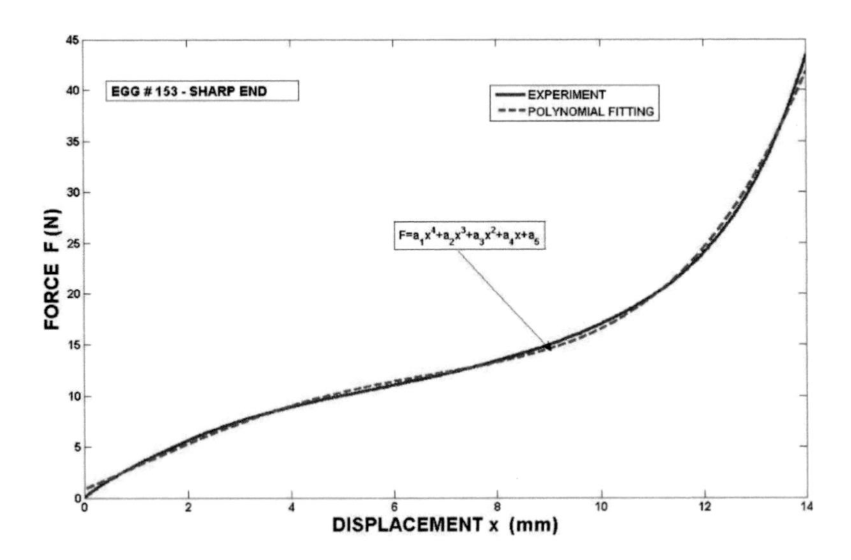

Figure 23. Stiffness of the egg loaded along the X – axis (sharp end).

These forces are given in the Tables 8 - 10. In these tables the values of the displacement at the fracture x_{max} and energy Q absorbed by the eggshell during the compression are also given. The values of Q are computed as

$$Q = \int_{0}^{x_{max}} F(x)dx \tag{15}$$

The distribution of these quantities is shown in the Figs. 24 – 26. The t-test of the given data revealed that there is a significant difference between values of the rupture function obtained for the egg loaded along X and Z axes. The rupture forces obtained for the egg loaded along the X-axis cannot be considered as different. The use of Pearson test led to the conclusion that the variability in the egg shape, eggshell thickness and, eggshell mass did not have any influence on the rupture force. The dependence of the rupture force on compression axes agree with results reported by Altuntas and Sekeroglu (2008). These authors also reported some other works supporting these results. The rupture force obtained at the compression test between two plates exhibited significant dependence on the egg shape.

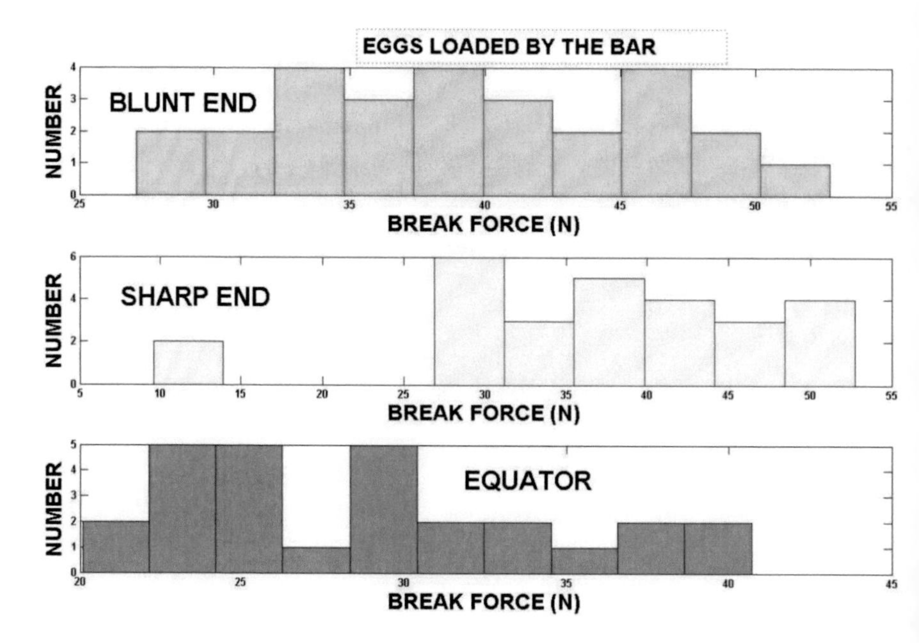

Figure 24. Distribution of the eggshell strength.

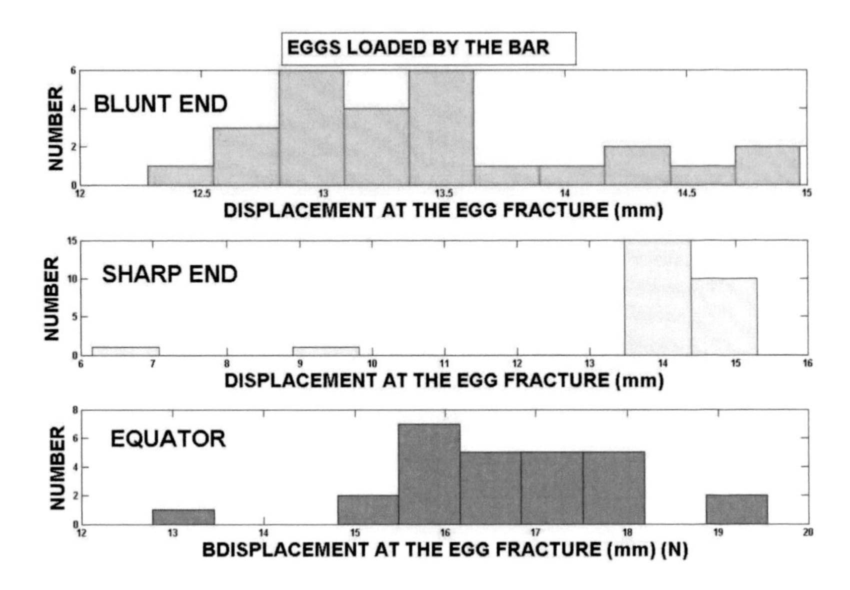

Figure 25. Distribution of the displacements at the eggshell break.

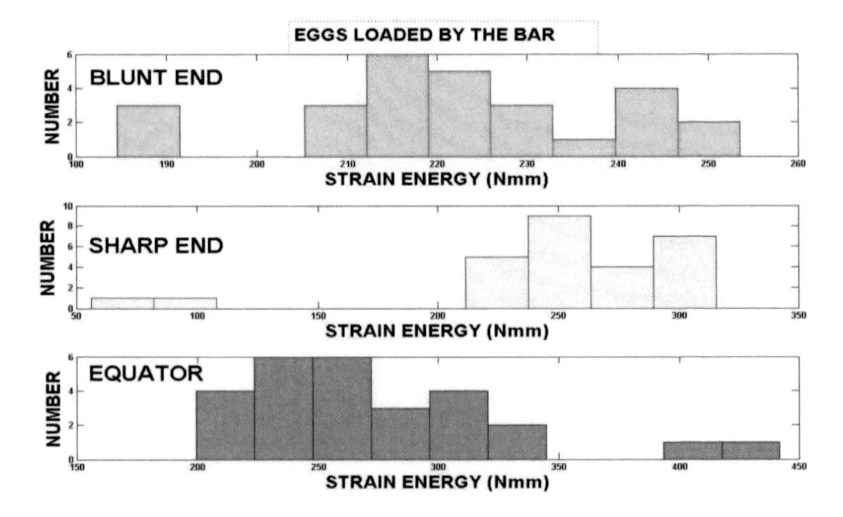

Figure 26. Distribution of the energy absorbed by the eggshell.

The independence reported in this chapter may be a consequence of the local character of loading. In order to verify these results, additional experiments on the eggs exhibiting higher variability in their shapes are needed. The obtained data serve namely for the comparison with data obtained at the impact loading.

Table 8. Static compression of the eggs – blunt end

Egg No.	Height (mm)	Width (mm)	Mass (g)	SI (%)	Strenght (N)	Displacement (mm)	Energy (Nmm)
151	57.6	44.4	64.0	77.08	52.16	17.23	500.88
152	55.8	44.0	61.2	78.85	30.45	14.04	262.95
153	56.7	42.5	57.5	74.96	46.72	14.21	311.13
154	60.4	46.1	72.0	76.32	36.70	14.08	274.46
155	57.4	43.5	61.7	75.78	30.70	13.51	236.54
156	60.9	46.0	72.3	75.53	42.31	14.96	298.73
157	58.2	44.4	65.2	76.29	49.30	15.01	315.39
159	58.0	44.5	64.1	76.72	33.98	13.79	247.56
160	59.3	43.9	64.6	74.03	29.93	13.59	232.49
161	56.8	44.8	65.6	78.87	26.90	13.57	222.56
162	57.2	43.2	61.0	75.52	36.52	14.20	261.18
163	61.7	44.2	67.7	71.64	9.57	6.16	56.16
164	58.9	46.0	71.2	78.10	35.96	15.24	279.92
165	62.0	45.3	72.3	73.06	30.43	13.63	240.79
166	57.2	44.4	63.0	77.62	42.63	14.75	278.04
167	56.0	41.9	56.1	74.82	41.42	14.03	254.23
168	56.4	43.9	61.7	77.84	33.30	13.66	240.53
169	63.5	45.2	74.5	71.18	51.21	14.40	295.92
170	59.3	44.9	67.3	75.72	41.91	14.55	271.73
171	56.8	43.7	61.1	76.94	52.76	14.53	291.33
172	59.3	45.0	68.2	75.89	39.75	14.75	262.67
173	58.4	43.8	62.8	75.00	34.54	13.66	234.50
174	54.4	42.4	55.5	77.94	36.34	14.10	244.45
175	57.3	44.1	62.9	76.96	48.85	15.30	291.56

Table 8. (Continued).

Egg No.	Height (mm)	Width (mm)	Mass (g)	SI (%)	Strenght (N)	Displacement (mm)	Energy (Nmm)
176	60.4	46.4	72.7	76.82	30.07	14.04	226.62
177	58.0	42.8	58.9	73.79	45.69	14.15	254.95
178	57.7	44.5	64.8	77.12	12.07	9.16	97.28
253	58.4	44.4	64.8	76.03	46.72	14.45	297.18
263	59.0	43.1	62.1	73.05	49.39	14.89	288.85
268	63.2	45.6	71.2	72.15	35.80	13.36	225.35
Average	58.5	44.3	64.9	75.72	37.80	13.90	259.86
std	2.10	1.10	4.95	1.98	10.18	1.84	68.79

Table 9. Static compression of the eggs – sharp end

Egg No	Height (mm)	Width (mm)	Mass (g)	SI (%)	Strenght (N)	Displacement (mm)	Energy (Nmm)
181	62.4	44.6	69.4	71.47	41.06	14.56	213.827
183	59.8	45.9	70.1	76.76	47.60	14.96	212.758
186	57.2	44.9	65.2	78.50	49.80	14.18	184.571
187	62.7	46.3	75.1	73.84	37.46	12.74	242.296
188	57.6	44.4	65.2	77.08	39.55	13.44	223.207
189	59.6	43.8	64.5	73.49	35.71	13.12	245.775
190	55.4	44.1	61.9	79.60	52.74	14.97	253.508
191	57.8	44.4	63.9	76.82	42.70	13.22	213.827
192	57.3	43.9	62.1	76.61	41.01	13.55	212.758
193	56.2	42.9	58.1	76.33	33.80	13.05	184.571
194	59.2	44.2	65.2	74.66	43.33	13.07	211.524

Egg No	Height (mm)	Width (mm)	Mass (g)	SI (%)	Strenght (N)	Displacement (mm)	Energy (Nmm)
195	59.4	43.9	65.9	73.91	46.07	13.24	229.463
196	56.4	45.6	66.2	80.85	49.17	14.10	232.312
197	55.8	43.0	58.8	77.06	32.61	13.53	236.303
198	59.7	44.9	67.6	75.21	39.28	12.91	208.455
199	61.4	46.4	74.5	75.57	40.56	13.42	231.407
200	59.2	43.7	64.5	73.82	39.66	12.86	216.893
201	59.7	44.6	66.3	74.71	33.39	12.85	213.827
202	57.6	45.0	66.1	78.13	27.12	12.71	212.758
203	57.7	43.8	62.2	75.91	36.54	12.28	184.571
204	58.1	44.2	63.3	76.08	46.54	12.60	211.524
206	58.0	44.3	65.2	76.38	45.66	13.68	242.296
207	57.6	42.9	60.6	74.48	29.44	13.19	223.207
208	58.2	45.2	67.7	77.66	36.63	14.66	245.775
209	56.9	45.6	66.7	80.14	30.07	14.37	253.508
244	59.4	46.1	71.4	77.61	33.21	13.08	220.627
249	57.7	45.3	66.8	78.51	31.71	13.55	223.140
254	60.3	45.2	69.7	74.96	42.40	13.55	221.423
259	57.4	43.0	60.8	74.91	35.42	13.47	231.337
264	57.4	44.6	64.5	77.70	35.10	13.27	219.914
Average	58.4	44.6	65.7	76.29	39.2	13.5	221.912
std	1.7	1.0	3.9	2.05	6.32	0.67	17.580

Table 10. Static compression of the eggs – equator

Egg No	Height (mm)	Width (mm)	Mass (g)	SI (%)	Strenght (N)	Displacement (mm)	Energy (Nmm)
245	57.5	44.9	65.3	78.09	28.16	16.91	259.548
255	59.4	45.7	69.2	76.94	25.33	16.15	248.346
260	56.3	42.9	58.7	76.20	33.62	17.90	302.310
269	61.4	47.8	78.4	77.85	38.11	12.78	341.391
271	57.7	43.7	61.6	75.74	31.86	19.54	441.994
272	59.1	44.8	67.1	75.80	38.97	19.16	396.437
274	58.8	44.2	64.8	75.17	30.25	17.59	306.032
275	59.8	45.3	69.2	75.75	29.82	17.24	298.157
276	59.2	45.6	69.4	77.03	31.73	17.26	293.975
277	59.0	46.3	71.2	78.47	33.33	17.99	312.254
278	56.0	45.0	63.8	80.36	28.79	17.06	267.554
281	57.8	42.7	59.4	73.88	24.76	16.46	249.745
282	60.7	44.4	67.8	73.15	23.10	16.48	249.323
284	55.6	43.3	58.6	77.88	23.84	16.14	225.622
286	61.6	44.9	69.1	72.89	22.61	15.99	236.632
287	56.7	45.5	67.1	80.25	28.38	15.76	221.703
289	60.6	47.4	75.4	78.22	34.54	16.98	267.447
291	58.7	45.4	68.3	77.34	40.70	17.51	292.045
292	59.4	45.4	69.7	76.43	24.36	18.11	325.213
293	56.7	44.7	63.0	78.84	27.19	16.04	225.622
294	58.7	46.6	70.8	79.39	26.25	17.55	284.695
295	58.7	46.4	71.4	79.05	22.94	16.32	241.275
296	57.9	44.8	66.0	77.37	29.91	15.70	218.574
297	58.1	44.8	66.8	77.11	20.85	16.70	263.377

Egg No	Height (mm)	Width (mm)	Mass (g)	SI (%)	Strenght (N)	Displacement (mm)	Energy (Nmm)
298	56.1	43.7	60.3	77.90	20.09	15.21	199.260
299	61.1	46.6	73.2	76.27	25.55	14.91	203.383
300	59.5	45.6	68.0	76.64	22.88	15.79	233.454
325	62.8	45.9	75.3	73.09	37.93	16.20	234.750
326	55.2	44.4	61.8	80.43	39.53	18.21	314.900
329	63.1	46.5	76.4	73.69	29.80	18.17	337.841
Average	58.8	45.2	67.6	76.91	29.17	16.79	276.429
std	2.0	1.2	5.1	2.06	5.67	1.29	53.773

3.2.2. Eggs Loaded by yhe Rod Impact

The main geometric characteristics of the eggs which have been tested by the procedure shown in the Figure 21 are given in the Table 11. Eggs have been tested by the rod impact in the three directions shown in the Figure 27.

Table 11. Geometry of the tested eggs (W – width, L – length, m – mass)

Egg No	L (mm)	W(mm)	m (g)	SI (%)
331	58.70	43.80	63.60	74.62
332	58.00	44.40	65.20	76.55
333	58.10	44.40	66.50	76.42
334	58.80	43.70	63.90	74.32
335	60.40	44.20	66.50	73.18
336	58.50	42.40	60.50	72.48
337	59.60	43.30	64.00	72.65
338	60.50	44.90	67.90	74.21
339	63.70	44.70	71.50	70.17
340	57.20	45.00	63.60	78.67
341	61.30	45.50	71.40	74.23
342	57.20	45.70	66.90	79.90
343	58.40	45.90	70.00	78.60
344	58.00	44.30	63.70	76.38
345	61.00	45.40	70.50	74.43
346	57.90	45.20	65.10	78.07
347	58.40	43.80	62.40	75.00
348	58.20	45.40	67.80	78.01
349	61.40	46.50	74.60	75.73
350	60.40	44.80	67.70	74.17
351	56.30	44.30	61.60	78.69
352	60.00	44.40	66.00	74.00
353	59.30	43.70	62.80	73.69
354	58.80	43.50	63.60	73.98
355	58.60	43.40	62.20	74.06
356	58.00	44.90	65.80	77.41
357	58.40	45.10	65.70	77.23
358	56.70	45.00	64.20	79.37
359	58.80	43.80	63.90	74.49
360	58.30	43.10	60.50	73.93

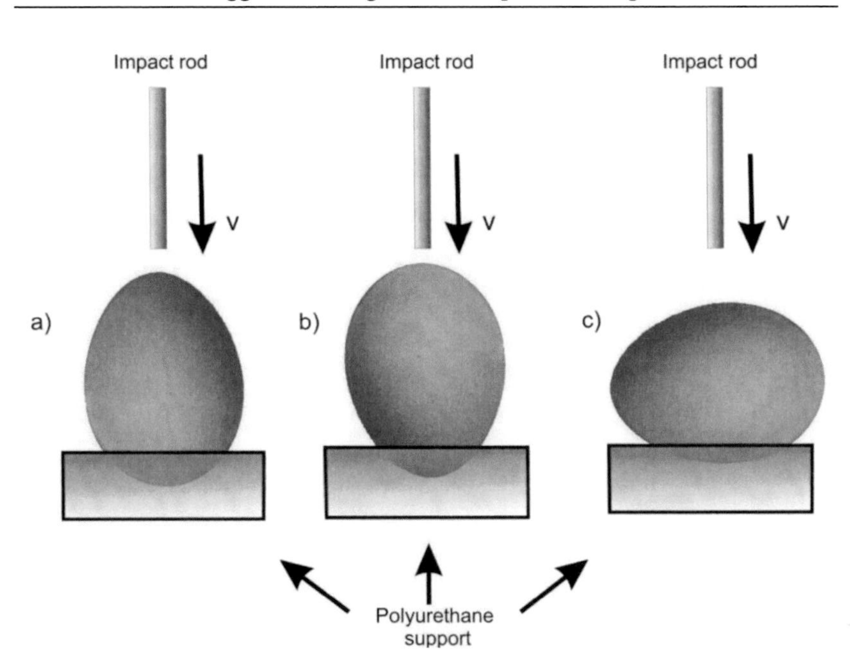

Figure 27. Schematic of the eggs loading.

In Figure 28 an example of the experimental record of the forces at the point of contact between bar and egg is shown. It has been found that the shape of the force-time function reflects the eggshell damage. If the eggshell is not damaged the shape of this function is nearly „half-sine“. The origin of the eggshell damage is connected with an abrupt in this dependence. The dependence of the force maximum on the height of the fall of the rod is shown in Figure 29. The time history of the eggshell surface displacement is shown in Figure 30. The origin of the damage leads to significant increase in peak value of this displacement. The same qualitative features, both force and surface displacement, have been observed for the remaining eggs.

The force corresponding to the eggshell strength has been chosen as the average between the force at the intersection of two lines (see Figure 29) and the highest value of the force at which no eggshell damage has been detected. The exact evaluation of the eggshell strength should be determined using the continuous increasing height of the rod fall.

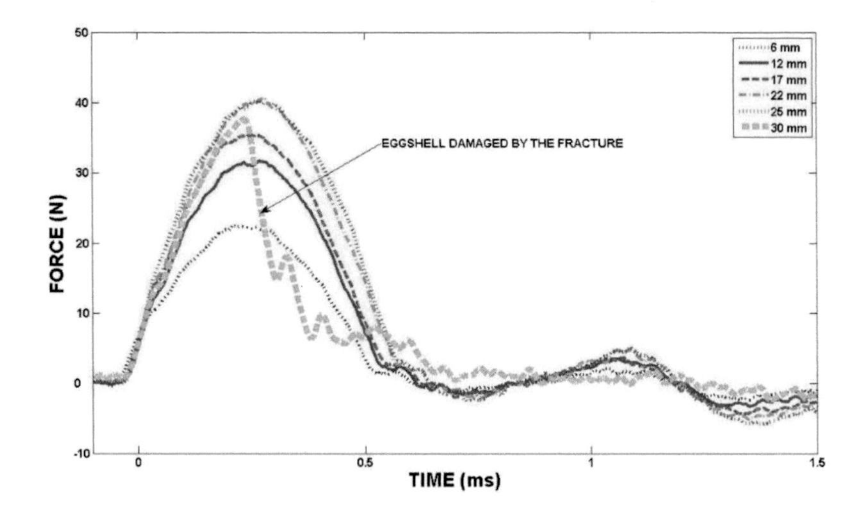

Figure 28. Experimental records of the time history of the force at the rod impact. Egg No. 331 – Rod impacted the blunt end of the egg. The displacement of the egg surface has been detected on the meridian at the distance of 36 mm from the impacted end. The different values of heights h are given in the upper right corner.

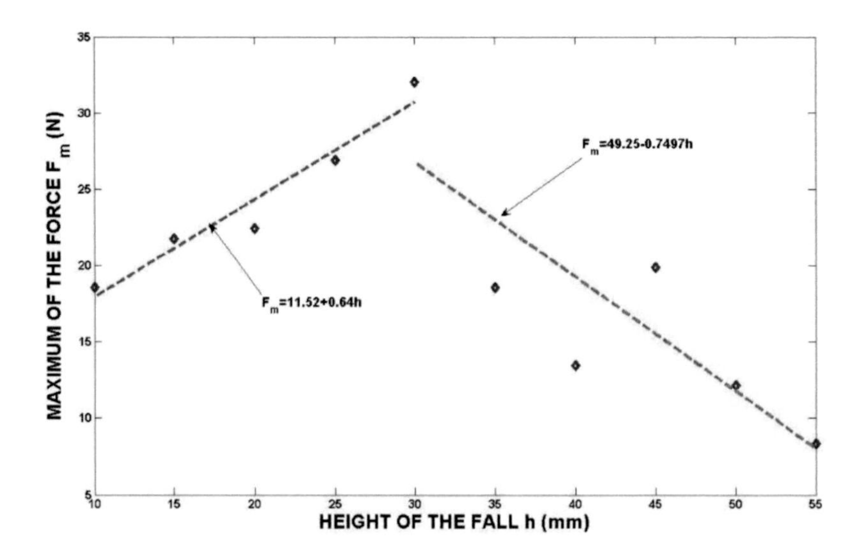

Figure 29. The dependence of the maximum of the force on the height of the rod fall.

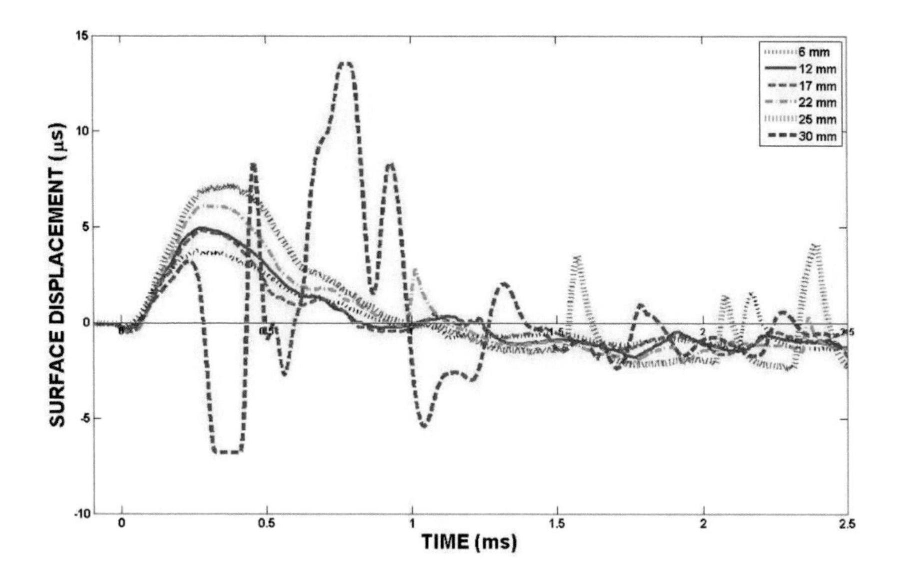

Figure 30. Surface displacement – time curves. Egg No. 331.

Experimental results can be also evaluated in the frequency domain using the Fast Fourier Transform – see chapter 2. The example of the amplitude of the spectral function of the force is shown in the Figure 31.

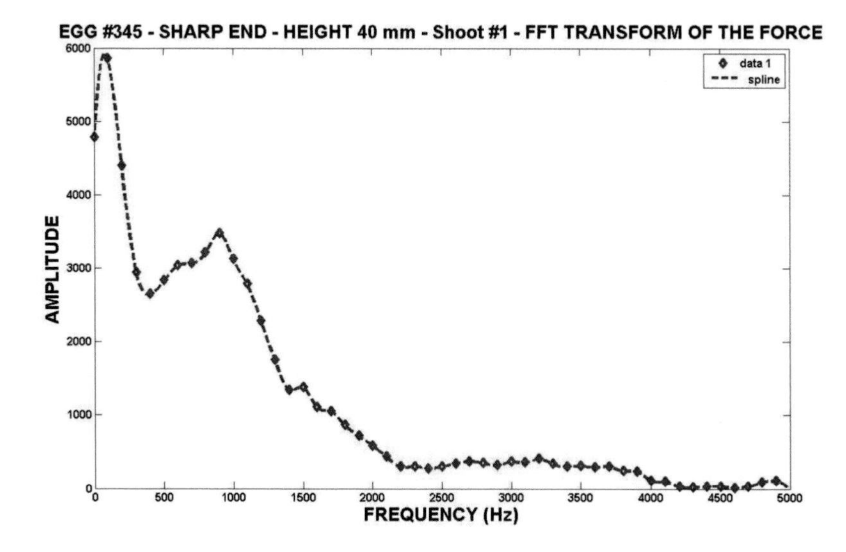

Figure 31. Frequency dependence of the spectral function.

Example of the spectral function for the displacement is shown in the Figure 32.

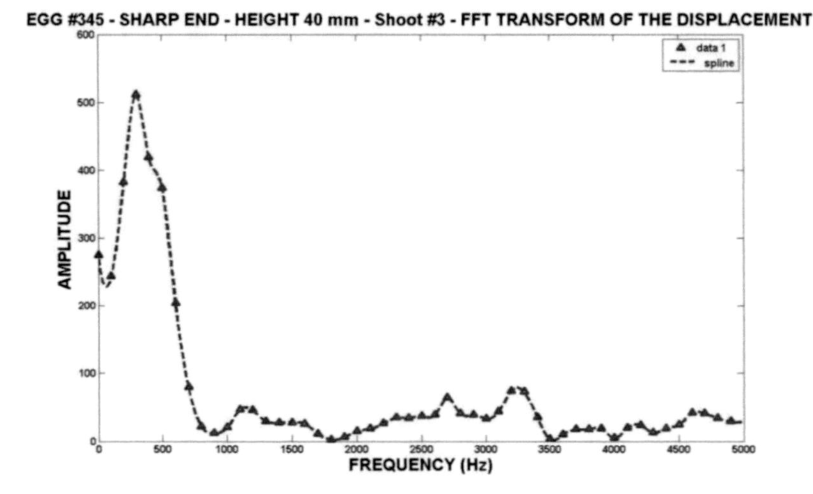

Figure 32. Frequency dependence of the spectral function.

The knowledge of the spectral functions both, force and amplitude, enables designing of the transfer function. Example of this function is given in Figure 33.

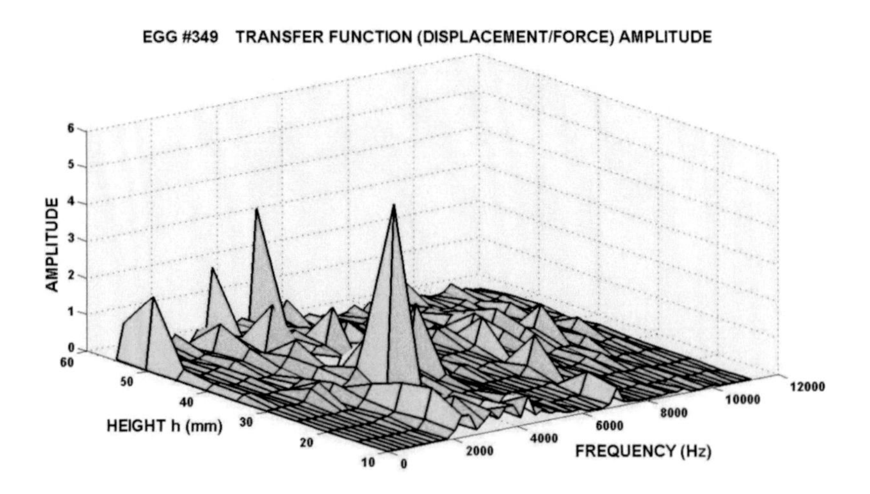

Figure 33. Amplitude of the transfer function.

There are some frequencies which are characterized by an amplification of this amplitude. The frequency dependence of the transfer amplitude for the eggs without any damage is presented in the Figure 34 there.

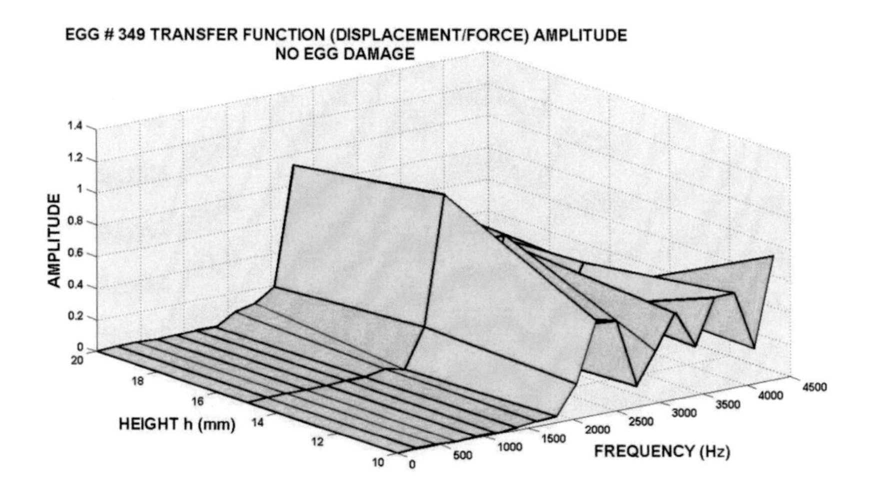

Figure 34. Amplitude of the transfer function.

The influence of the egg damage is shown in the Figure 35.

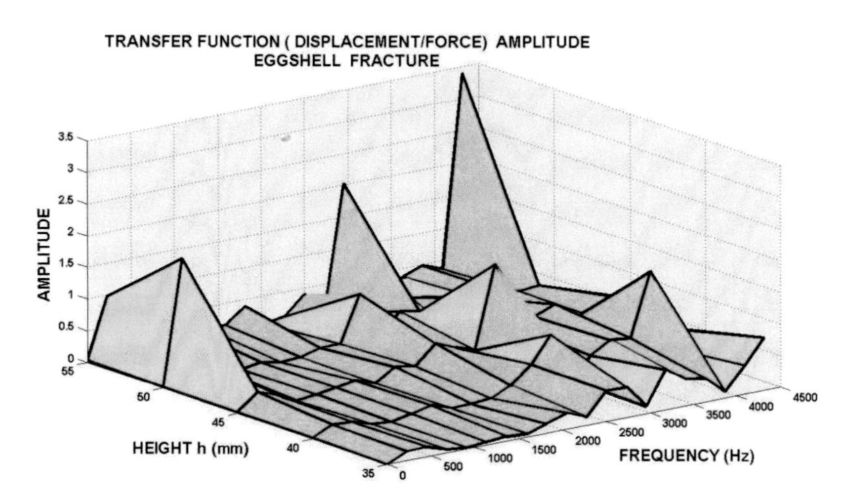

Figure 35. Amplitude of the transfer function.

Mean values of the transfer function amplitude are displayed in Figure 36.

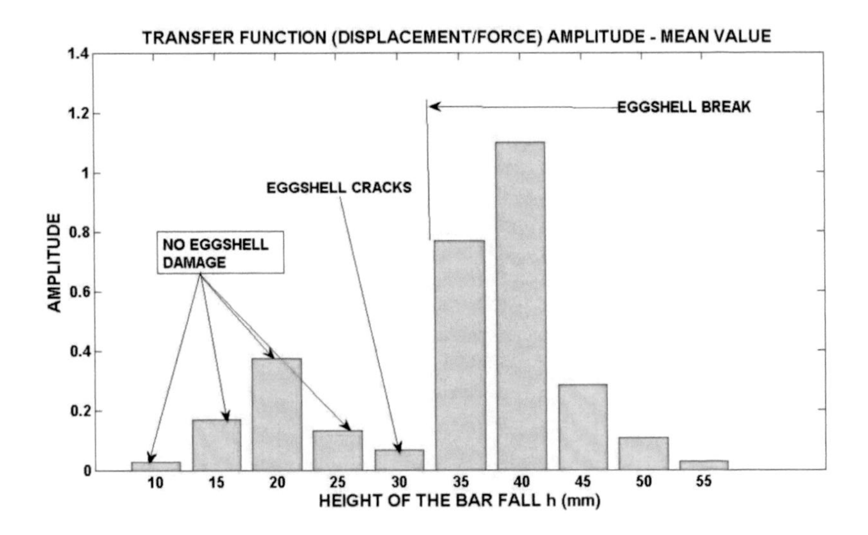

Figure 36. Mean values of the transfer function amplitude.

One of the objections to this method can consists in the possible initiation of the micro-defects during subsequent loading of the eggshell. Such micro-cracks can seriously affect the strength. In order to study this problem, an experiment with repeating loading of the egg has been performed. The height of the rod fall has been constant (h = 22 mm). Ten repeated impacts have been performed. The eggshell has not exhibited any visible damage. The shape of the force - time curve is typical for non-destructive impact. The maximum values of the force are presented in Figure 37. It can be seen that these values are very close to the strength of the egg. There is an exception at the 6th and 7th impacts. This effect may be a consequence of certain changes of the impact conditions (lower height). In this figure the higher amplitudes of the impact force are also plotted.

The values of the eggshell strength are shown in Figure 38. Ten eggs have been used for the evaluation of the dynamic strength for all three impact orientations (sharp end, blunt end, and equator). The basic statistics of the given data is given in the Table 12 and Table 13. The maximum of the dynamic strength can be observed for the impact of the rod on the sharp end. The minimum value of strength can be found for the rod impact on the equator.

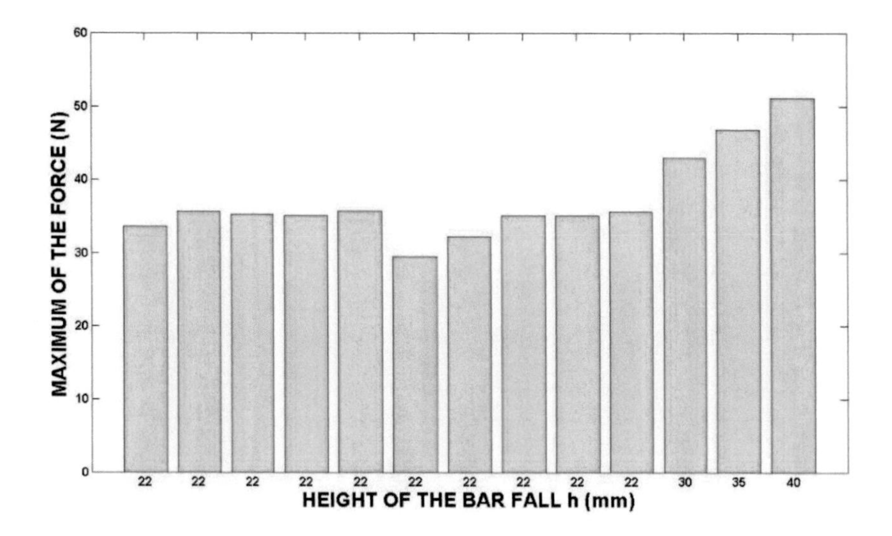

Figure 37. Maximum of the force during rod impact. Egg No. 339. Height of the fall h = 22 mm. Rod impacted the blunt end of the egg.

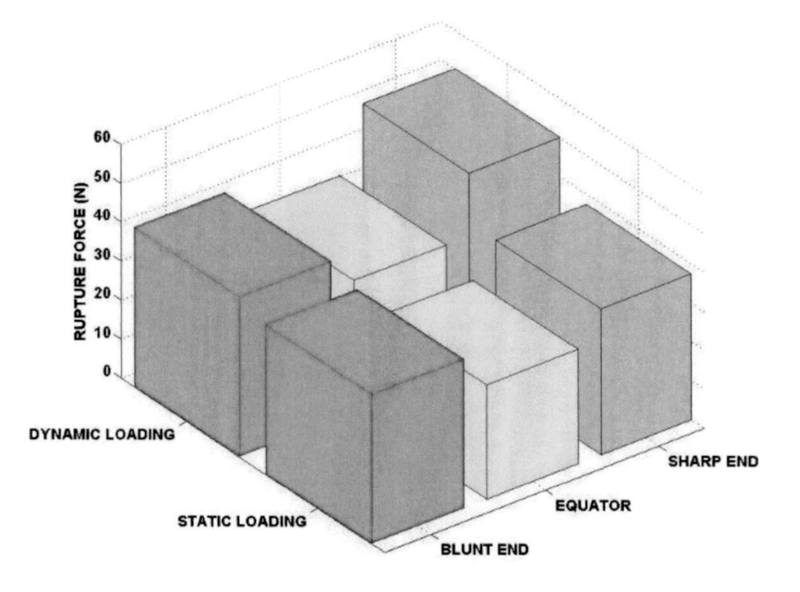

Figure 38. Dynamic and static strength of the eggshells.

These results qualitatively agree with the results obtained at the static loading. Contrary to the results of the static loading tests, there is a significant difference between the rupture force at the sharp and blunt end. The results also suggest that dynamic rupture force is higher than that obtained at the static loading. This can be taken as an evidence of the influence of the loading rate. This phenomenon has been reported in some papers - see e.g. Altuntas and Sekeroglu (2008). In order to obtain a more detailed insight into this problem some other experiments are needed. It is particularly necessary to obtain results for different speeds of the rod at the static loading.

Table 12. Average values of the static rupture force F_r

Point of loading	Rupture force F_r (N)	Standard deviation (N)
Blunt end	38.2	± 7.9
Equator	29.3	± 6.2
Sharp end	37.4	± 10.2

Table 13. Average values of the dynamic rupture force F_r

Point of the rod impact	Rupture force F_r (N)	Standard deviation (N)
Blunt end	40.9	± 3.0
Equator	33.9	± 3.6
Sharp end	50.1	± 5.2

As it has been mentioned the values of the eggshell strength in terms of the rupture force are affected by many factors (egg specific gravity, egg mass, egg volume, egg surface area, egg thickness, shell weight, shape and shell percentage). In order to avoid the simultaneous influence of these factors the numerical simulation of these experiments should be performed. Some preliminary results in this field are involved in the following paragraph.

3.2.3. Numerical Simulation of the Rod Impact

For the numerical simulation of the experiments described in the previous paragraph, the LS DYNA 3D finite element code has been used, as in the case of simulation of the ball impact. The same model of the egg has been used.

The numerical model of the experiment is shown in Figure 39. The following problems have been solved:

Impact of the rod on the blunt end of the egg (Egg 331 has been used)
Impact on the egg equator (Egg 349)
Impact on the sharp end of the egg (Egg 347)

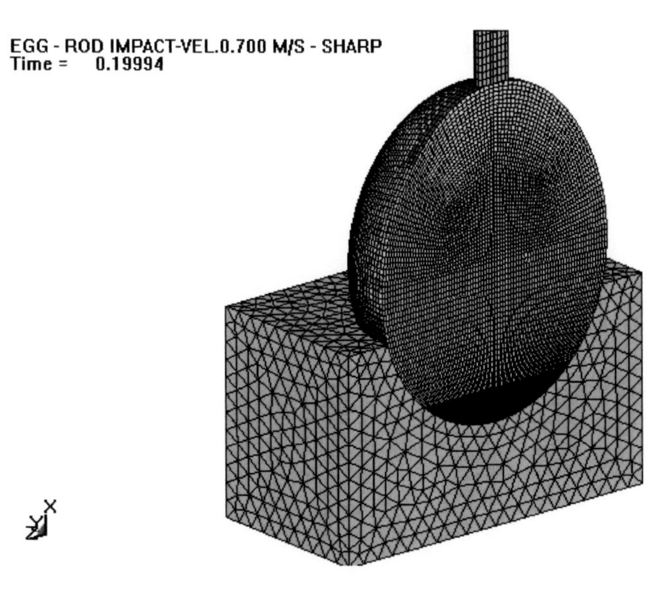

Figure 39. Finite element model of the solved problem.

Table 14. Main characteristics of the eggs used in the numerical simulation of the impact loading

Egg No.	Mass m (g)	Shape index (%)	E (GPa)	Eggshell thickness (mm)
331	63.6	74.62	73	0.230
347	61.4	75.73	47	0.235
349	62.4	75.00	47	0.240
339	71.5	70.17	45	0.235

The main characteristics of the used eggs are given in the Table 14. The height of the rod fall has been kept at 25 mm. The impact velocity of the rod is than 0.7 m/s. In order to verify the validity of the model of the egg, the time histories of the forces and surface displacements have been evaluated. These data can be compared with experimental ones. In Figs. 40 and 41, the numerical and experimental records obtained for the egg No. 331 are presented. Results for the remaining problems exhibit the same qualitative features. The computed peak values of the force agree well with experimental ones. There are some differences in the time course of both functions. Owing to some problems with the force record (finite length of the strain gauges etc.) the observed discrepancy seems be acceptable. Very reasonable agreement has been exhibited between computed and experimentally recorded time histories of the surface displacements.

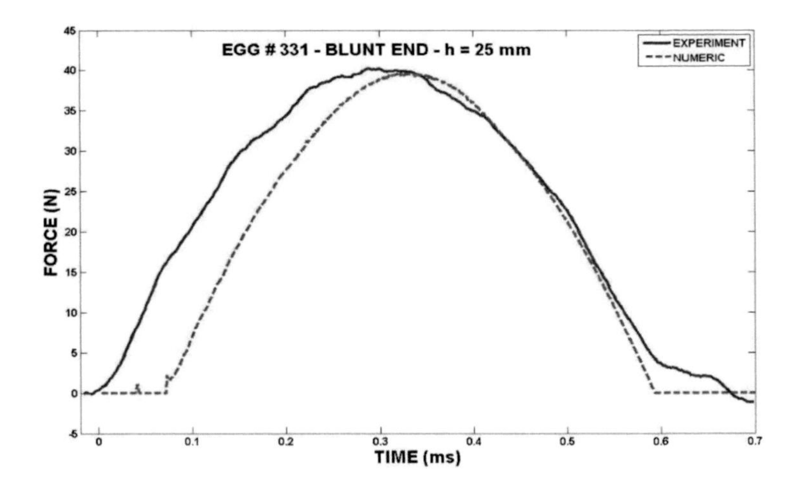

Figure 40. Experimental and numerical time histories of the force at the contact between the rod and the eggshell.

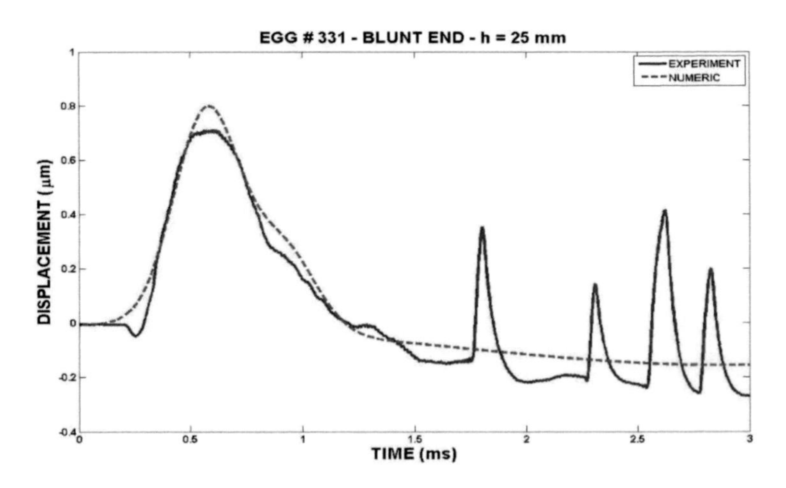

Figure 41. The dependence of the surface displacement on time. Comparison between numerical and experimental results.

The numerical results did not respect some experimentally observed peaks. These peaks are probably a consequence of possible transient phenomena in the recording system. Their occurrence has been detected only for a limited number of experiments. If we take into account some assumptions, namely the assumption on the egg liquids behavior, the agreement between numerical simulation and experiment seems to be more than satisfactory. Owing to this fact the next results of the numerical computations can be accepted as relatively reliable. The development of the stress state after the rod impact is documented in the Figure 42a-c.

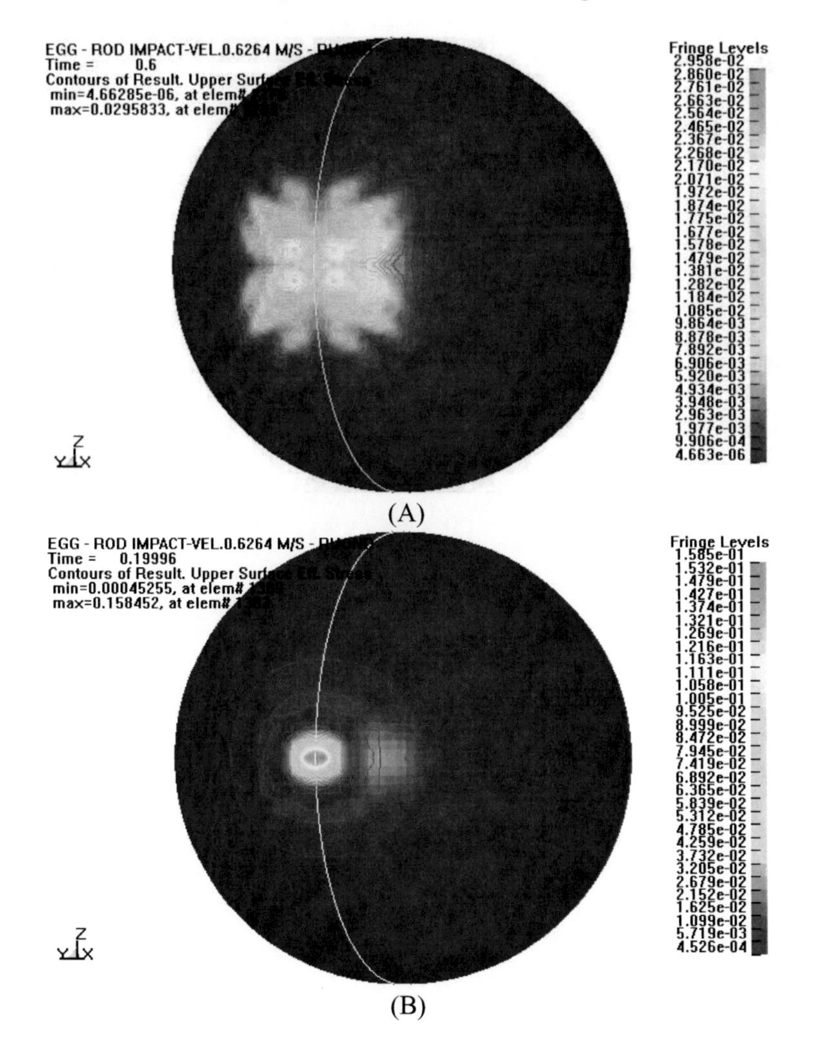

(A)

(B)

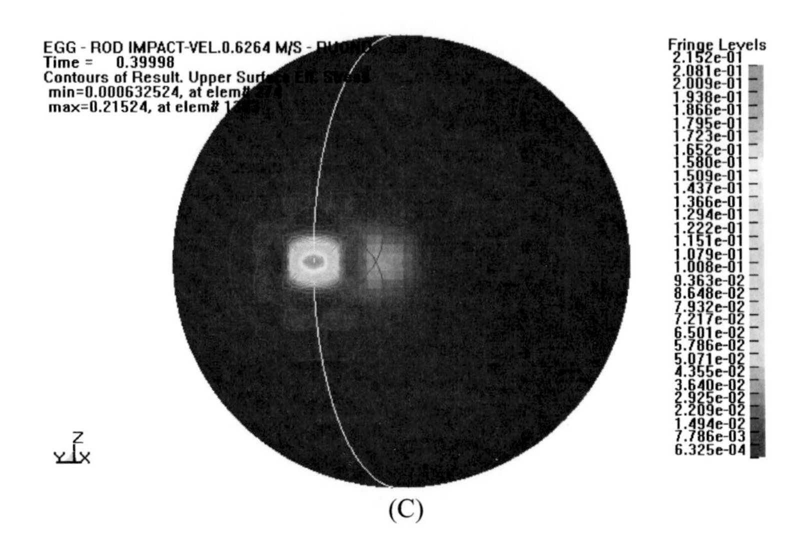

(C)

Figure 42 a-c. The development of the equivalent stress in the eggshell. Egg No. 331 - blunt end. a) time = 0.19996 ms; b) time = 0.39998 ms; c) time = 0.6 ms.

In these figures the time development of the equivalent stress is displayed. It can be seen that the stress is localized on a relatively small area around point of the contact between rod and eggshell. In the Figure 43 the computed displacements at the point of impact are shown.

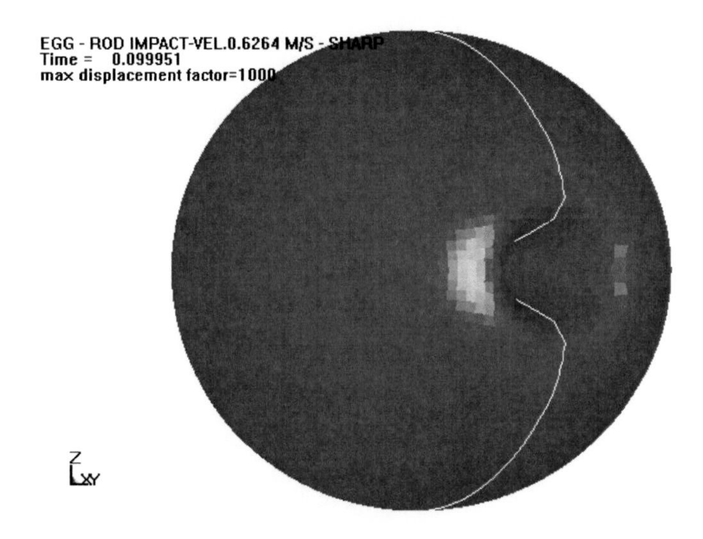

Figure 43. Computed displacement. Impact on the sharp end. Impact velocity: 0.6264 m/s.

In the next step the stress development in the eggshells has been modeled. This computation has been performed for the impact conditions at which the eggshell rupture has been observed (impact velocity = 0.741 m/s). The distribution of stresses through the eggshell thickness is the most appropriately examined using the solid element model. Directly under the rod impact on the inner surface of the shell, an equi-biaxial stress distribution develops in which the hoop and meridional stress are equal. The development of the stress on the outer and inner surfaces of the eggshell is shown in Figs. 44a, b. More detailed analysis revealed that the compressive stress decreases with the distance from the point of the rod impact and it changes to the tensile stress. The stress at the inner surface is only tensile. Very similar features of the stress distribution in the eggshells have been also reported for the numerical simulation of the quasi-static compressive loading of the eggs - see MacLeod et al. (2006). If we take the maximum of the tensile stress as a measure of the eggshell strength, we obtain results given in the Table 14. It seems that these stresses are independent on the position of the point of the rod impact.

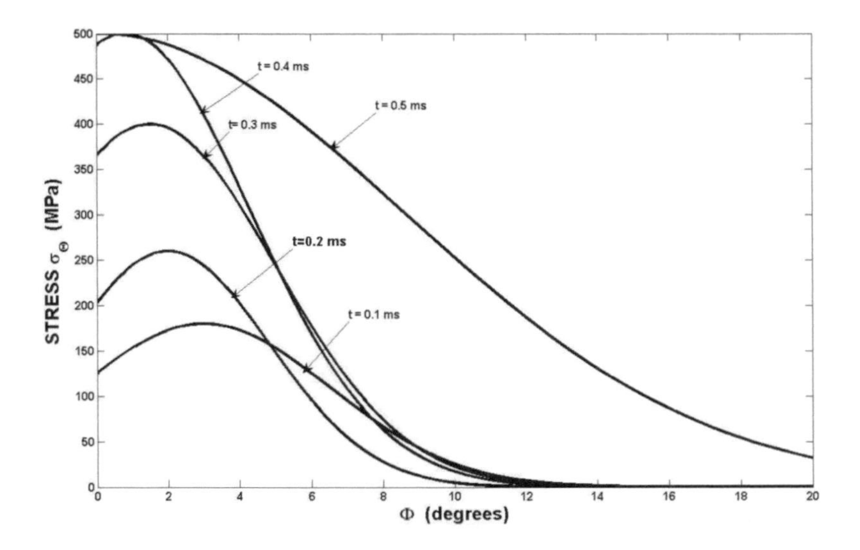

Figure 44a. Time development of the stress at the inner surface of the eggshell.

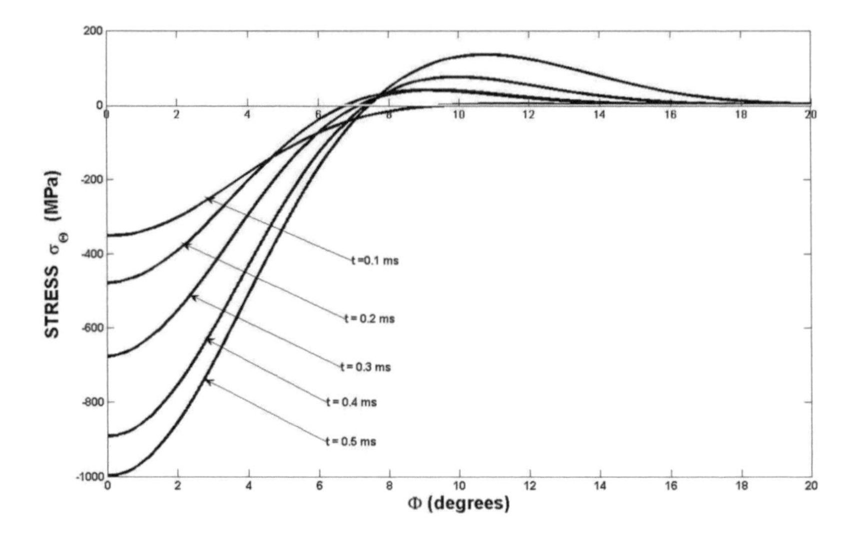

Figure 44b. Time development of the stress at the inner surface of the eggshell.

Although there is still a lack of experimental and namely the numerical results, the obtained results suggest that the value of the maximum of the tensile stress at which the eggshell rupture starts can be independent of many factors affecting the value of the experimentally found rupture force. This stress can be considered as an intrinsic strength property of the eggshell material. In order to verify this hypothesis many other experiments are needed.

Chapter 4

HEN'S EGGS IMPACT AGAINST TO ELASTIC OBSTACLE

Breakage occurs whenever the stress at some point in the shell exceeds the ultimate strength of shell material at given point. When attempting to predict when breakage may occur two questions arise:

- What is the maximum external force upon the shell?
- What is the maximum consequential stress within the shell?

These problems must be solved for all kinds of eggs loading. When breakage occurs under dynamic conditions question *(a)* is concerned with ballistics and the elasticity of shell material. The ballistic aspects are complicated by two factors: the shape of the egg and the nature of its contents. At most points on the shell of an egg-shaped body a line drawn normal to the surface does not pass through the centre of mass, so any impact results in rotation as well as translatory movement. Because the egg contents include yolk and thick white floating in thin white (but moored to the poles by chalazae), rotation about the long axis may involve little more than the shell, whereas rotation about a transverse axis must involve the greater part of the egg. The distribution of the energy acquired at impact into three components corresponding with translation, rotation about the long axis and rotation about a transverse axis cannot be predicted accurately without further information about the properties of the egg contents, but it seems likely that in all situations where shell breakage may occur the energy of rotation about the long axis will be small compared with the sum of the other two components. It is therefore disregarded in what follows. Conversely, for rotation about a transverse axis the egg contents are treated as though they were solidly anchored to the shell. Given these assumptions and the further assumption that eggs behave as elastic bodies, their change of motion at impact can be predicted from their motions before impact and their masses and moments of inertia.

Question *(b)* relates the external force, *P,* on the shell to the location and magnitude of the maximum stress, *Stm,* within it. This problem has been solved namely for the eggs compressed between two flat plates at quasistatic loading conditions. The shell theory of engineers, as developed for shallow spherical domes (Reissner, 1946) and for flat plates loaded by a concentrated force (Timoshenko and Woinowsky-Krieger, 1959), has been applied to the case of an egg shell loaded at its poles. The parameters of the theoretical model shell are its thickness, *T,* and radius, *R,* the radius, *r,* of the area over which it is loaded and Young's Modulus, *E,* Poisson's Ratio, *ν,* and the ultimate tensile strength, *Stu,* of the material of which the shell is made. With this model the stress is expected to be maximal at the inner surface of the shell below the centre of the loaded area and the expected values of *Pm* can be calculated in terms of *T, R, r, E, ν* and *Stu.* However, the values of *Pm* observed experimentally were greater, by a factor of 20, than predicted values based on the best available estimates of *E, ν* and *Stu* (Voisey and Hunt, 1967a). Later work in which eggs were compressed between flat plates (Tung et al., 1969) likewise led to estimates of failure stress that were consistent but considerably higher than those obtained previously by methods that measure membrane stresses (Hammerle and Mohsenin, 1967; Sluka et al., 1967). These findings call into question the suitability of the model. In one respect, at least, it is clearly unsuitable; it does not take into account the existence, discovered subsequently, of an inner layer of shell that is weak in tension (Carter, 1970, 1971). A quite different model, in which the shell is treated as a set of incompressible radial prisms held together by elastic material, leads to the conclusion that the stress at the inner surface of an area of shell flattened by an external load is independent of the load: increasing *P* merely increases the contact area, in such a way that *r/P* remains constant. Data of Brooks and Hale (1960) v show that this is the relationship between *r* and *P.* Since experience nevertheless teaches that the probability of shell fracture is related to *P,* this model implies that the maximal stress, and therefore the stress that initiates fracture, must occur not within the flattened area but at some point outside it. The photographs of Voisey and Hunt (1967a), who compressed eggs that had been covered with a strain detecting brittle coating, point to the same conclusion; the cracks that were formed first - the continuous "main roads" that are joined but not crossed by the "side roads" - do not run through the contact area but run outwards from points near it. This implies that they were formed as a result of circumferential tensile stresses at some distance from the contact area. In that case the relevant shell thickness is *Te,* the thickness that is effective in respect of tensile strength. The difference, ε, between *T* and *Te* depends on the strain and identity of the hen; strain mean values range from about 85 to 130 (Carter, 1970c, 1971). Time is another parameter that may have to be taken into account, since the external force that will just crack an egg shell has been shown experimentally by Voisey and Hunt (1969) to depend on compression speed, at least if the speed is below about 10 mm/s. As these workers also showed that the stiffness of an egg shell is nearly constant at such speeds, *Stu* must be a function of compression speed, at least for speeds up to about 10 mm/s. It is not clear whether or not *Stu* depends on compression speed at the much higher speeds, usually in excess of 200 mm/s, at which shell cracking may occur at impact. The possibility of such dependence must be borne in mind, though if the energy-absorbing capability of egg shell is limited, as the same workers suggested, the rate of change of *Stu* with impact speed may be low. A consistent feature of egg shell strength is its high variability under apparently similar experimental conditions. It is due in part to the fact that egg shell is brittle material, in the sense in which engineers use the term. However, this is not the whole explanation, since an appreciable part of the residual variation in egg shell strength, after overall shell thickness and curvatures have been taken into account, is usually associated with the identity of the hen. This suggests that between-hen variation in *Stu* or ε may play an important role.

The theories mentioned above have been used by Carter (1976) for the solution of many problems of the hen's egg impact. The theory predicts the maximum of the force exerted on an egg at impact. As an example, let us consider the impact of an egg against an obstacle with the mass and the stiffness which are much smaller than those of the egg. This example represents e.g. the impact of the egg against an elastic bar. The maximum force P_m is given by:

$$P_m = \left(\frac{1}{2}Mv_o^2\right)^{\frac{1}{2}} S^{\frac{1}{2}}, \qquad (16)$$

where M is the mass of the egg, v_o is its impact velocity and S is the eggshell stiffness. Although this theory exhibited a reasonable agreement with experimental data its use is limited only to some very simple impact conditions. The most effective way how to describe the impact problems consists in the use of the numerical simulation. This procedure has been very successful for the analysis of the quasi static compression of the eggs (MacLeod et al., 2006) and also for the explanation of some phenomena connected with the dynamic loading (Nedomová et al., 2009). The verification of the results of the numerical simulation must be based on the reliable experiments. These experiments are described in the following section.

4.1. Experimental Arrangement

The research of egg impact loading was performed with use of specially developed testing method – see schematic in Figure 45. The egg falls from selected height on the round section bar (diameter of 50 mm) made of PMMA. The egg is (during its fall) guided in the hollow cylinder, in order to prevent its swinging. The egg falls either on its sharp or blunt end. The impacted bar is deformed purely elastically. Thus determination of the force in the contact point between egg and bar is possible. The force is quantified by use of strain gauge attached to the bar.

4.2. Experimental Results

The hen's eggs were dropped from different heights ranging from 100 to 780 mm. The time dependencies of the force acting in the impact point were determined. In order to evaluate the influence of egg liquids on the impact behavior, the raw as well as boiled eggs were examined. The eggs were kept in boiling water for 10 minutes. The data were evaluated and processed in time as well as frequency domain. Regarding the response and sensitivity of force sensing, only those experiments were relevant, where the eggs were dropped from 105 (and more) mm. In this case, the breakage of the eggshell always occurred. The character of the breakage is shown on Figs 46a – d.

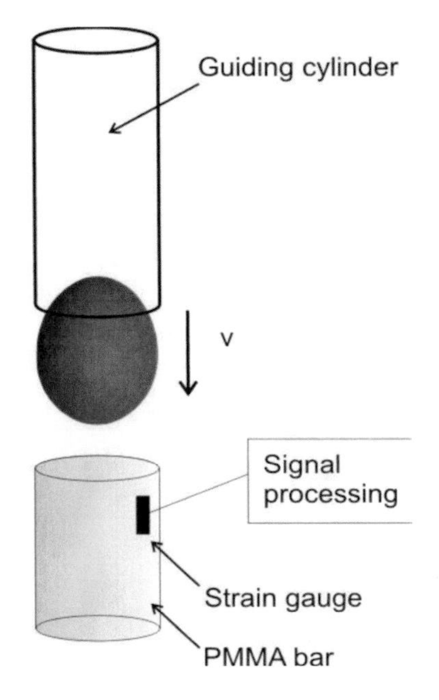

Figure 45. Schematic of the experimental arrangement.

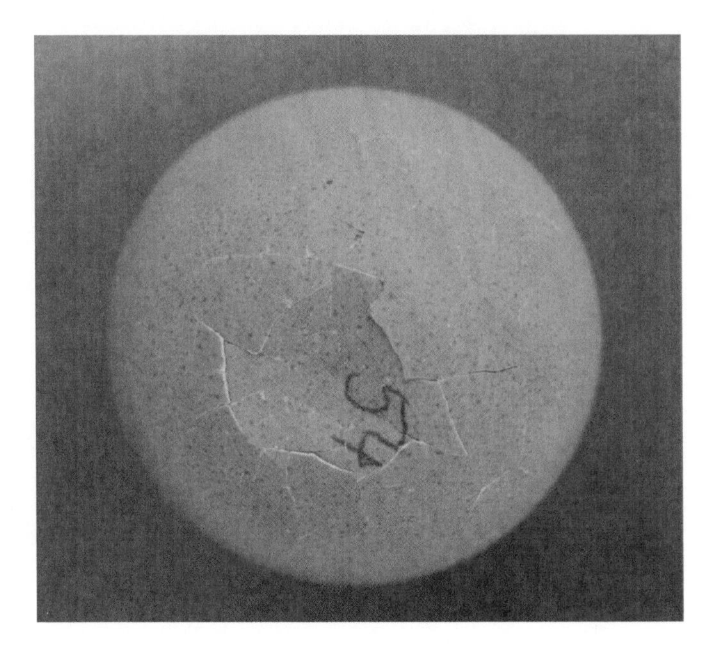

Figure 46a. Boiled eg, fall height 200 mm – blunt end.

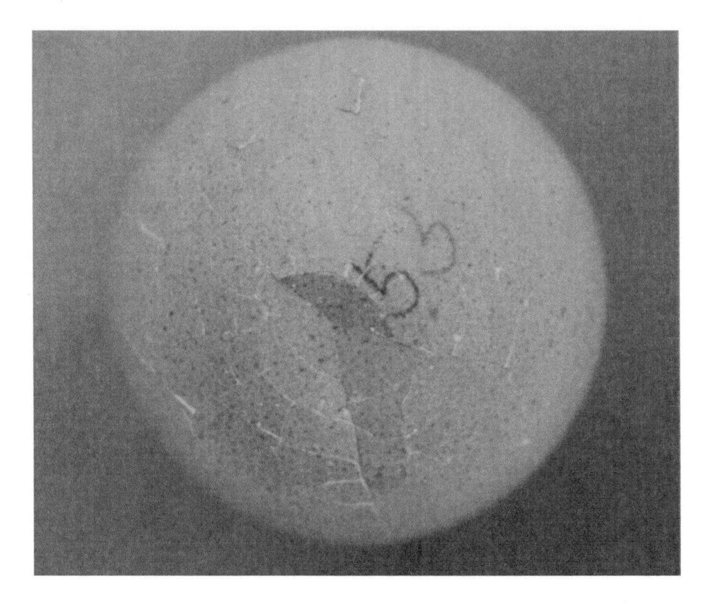

Figure 46b. Boiled egg, fall height 300 mm – blunt end.

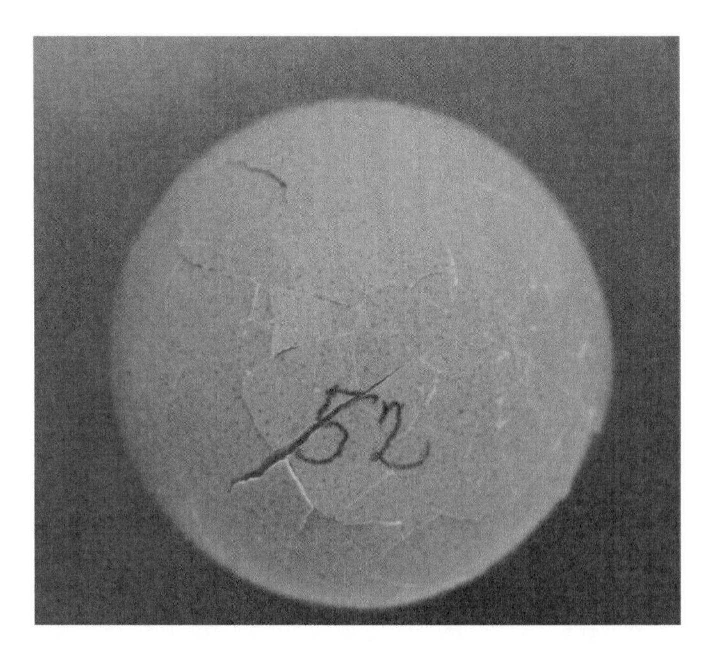

Figure 46c. Boiled egg, fall height 400 mm – blunt end.

Figure 46d. Boiled egg, fall height 500 mm – blunt end.

4.2.1. Raw Eggs

Time histories of the force acting in the point of egg-bar contact are presented in the Figs 47 – 52. The different fall heights (or impact velocities) are considered.

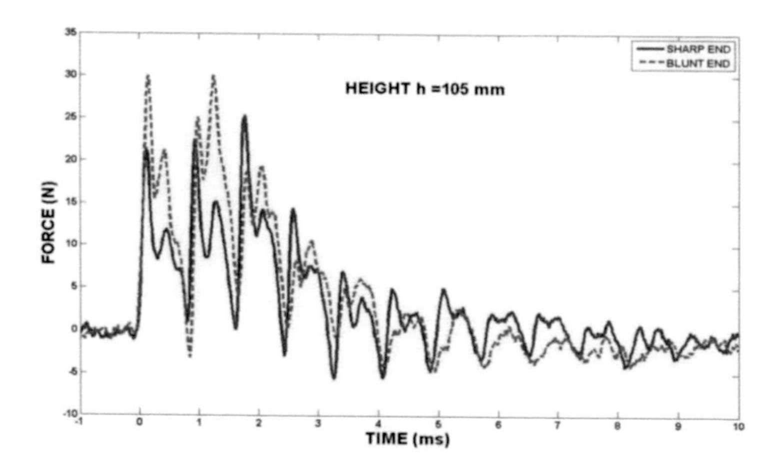

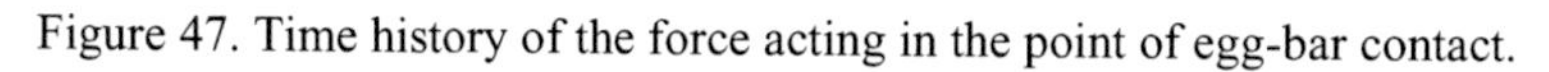
Figure 47. Time history of the force acting in the point of egg-bar contact.

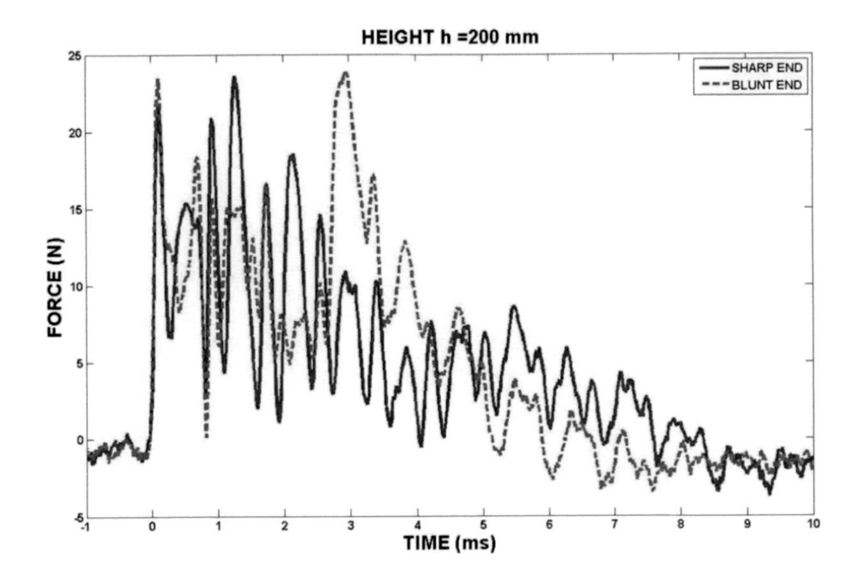

Figure 48. Time history of the force acting in the point of egg-bar contact.

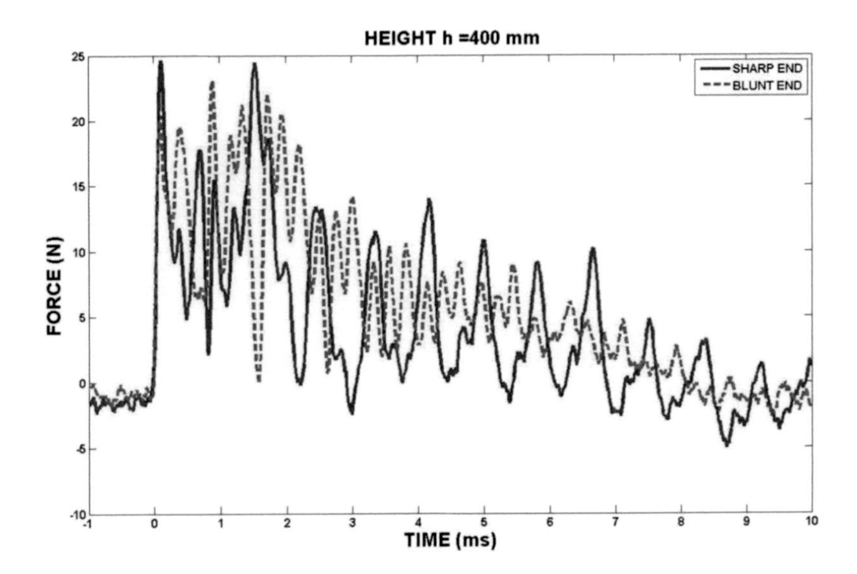

Figure 49. Time history of the force acting in the point of egg-bar contact.

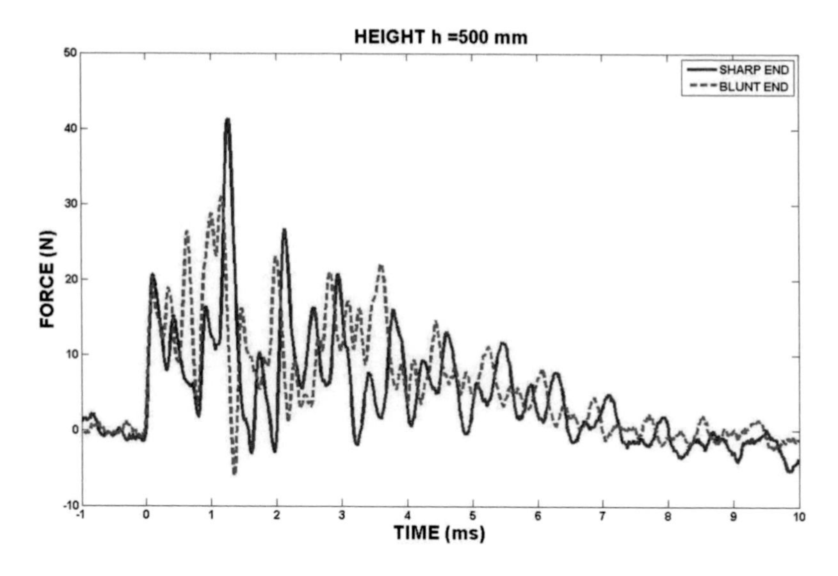

Figure 50. Time history of the force acting in the point of egg-bar contact.

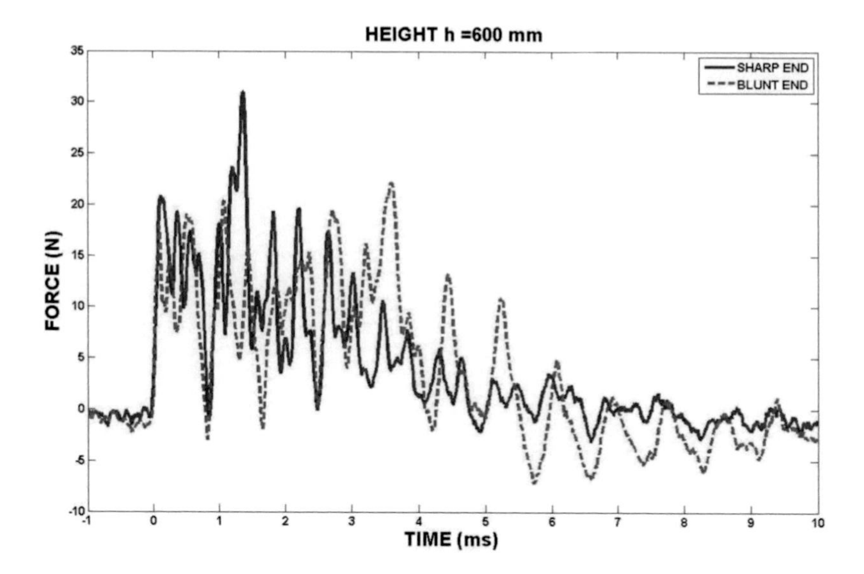

Figure 51. Time history of the force acting in the point of egg-bar contact.

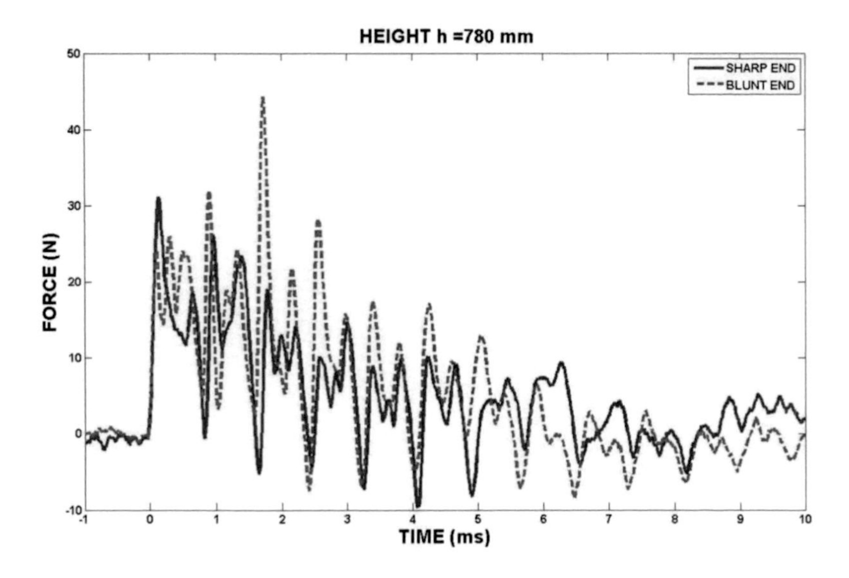

Figure 52. Time history of the force acting in the point of egg-bar contact.

The course of contact force is generally connected with numerous oscillations, which is valid for all fall heights and both egg ends. The amplitudes of individual oscillations are characterized by similar size. The values of maximum forces are visualized in Fig 53.

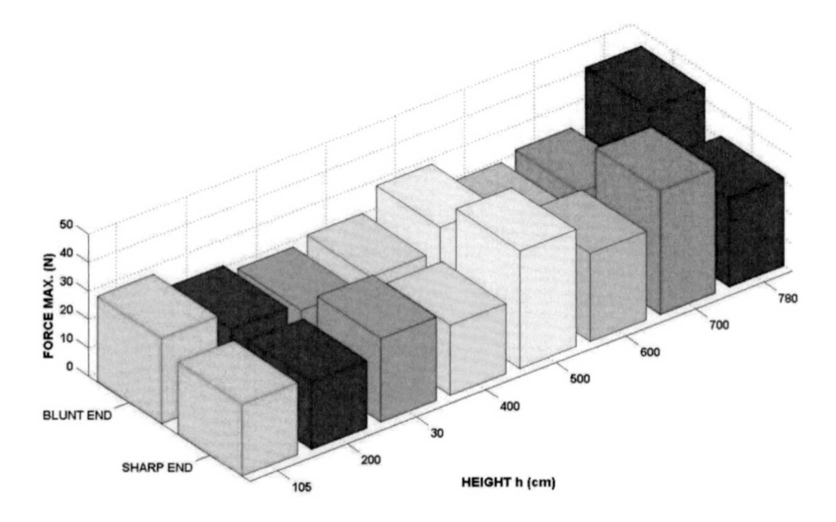

Figure 53. Maximum contact forces.

It is evident that maximum of the contact force is independent on impact velocity and it is not possible to find the relevant difference between force values during fall on either sharp or blunt end. These results are visualized in Figure 54.

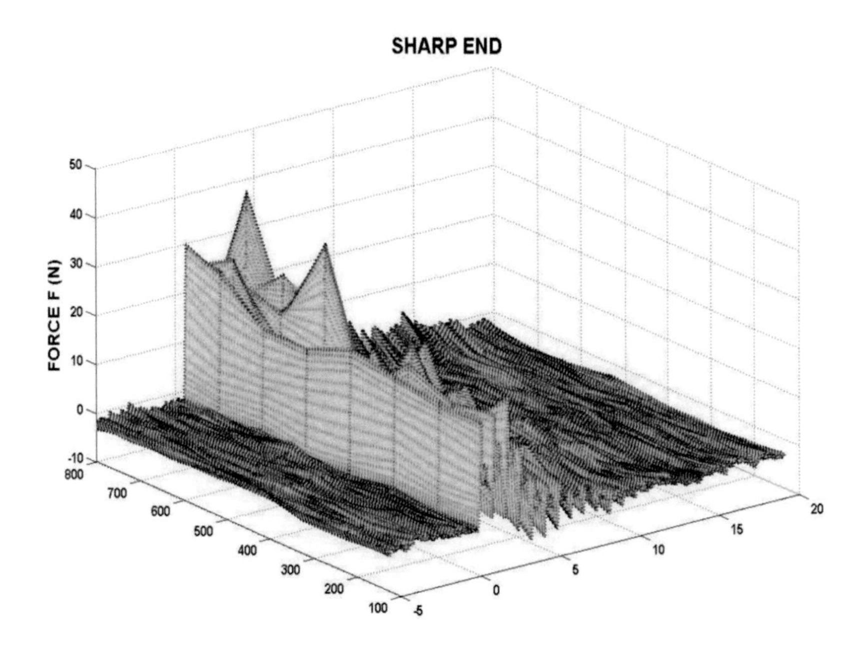

Figure 54. Influence of fall height on the force time history.

Considering rather complicated time history of the contact force *F(t)*, use of impulse of the force seems to be more efficient:

$$I = \int F(t)dt \tag{17}$$

The example of time history of this value is shown in Figure 55.

The decrease of the variable is connected with negative values of oscillating force. Fall height dependence of this variable is presented in Figs 56 and 57.

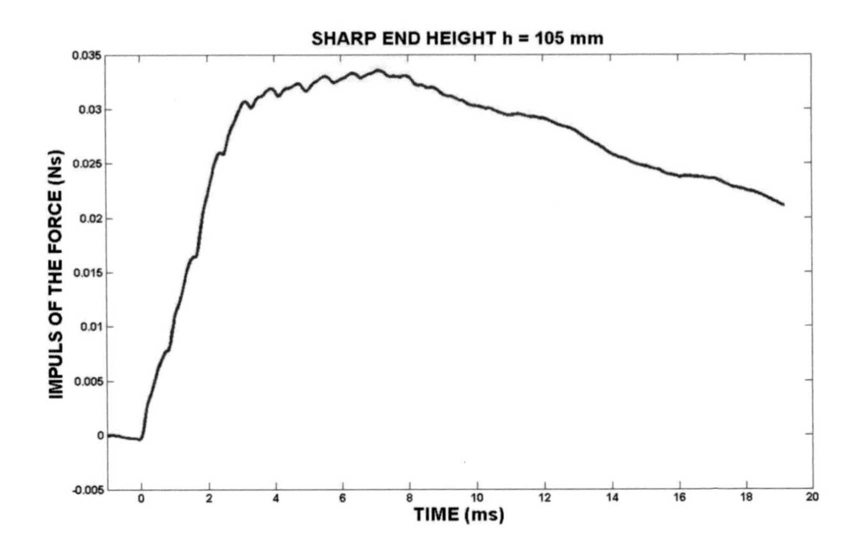

Figure 55. Time history of the impulse of the force.

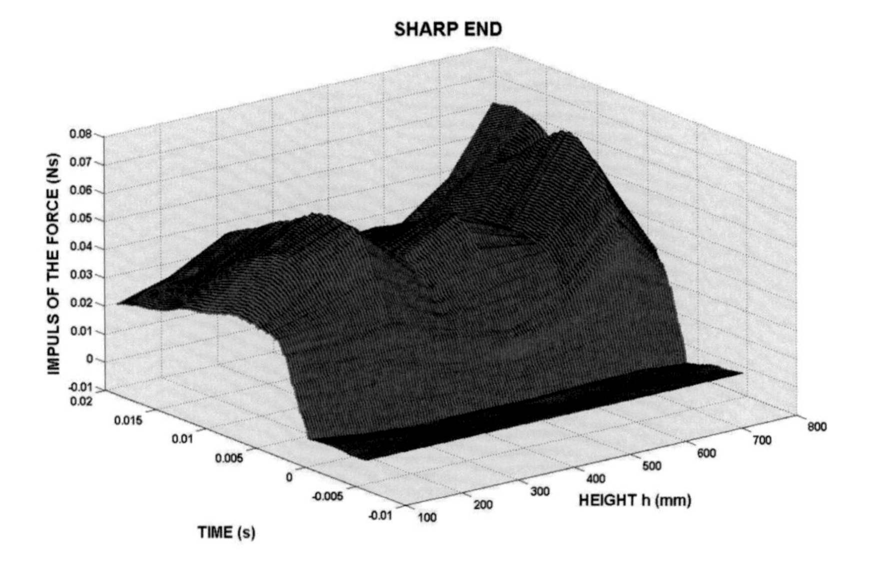

Figure 56. Influence of the fall height on the time history of the impulse of the force.

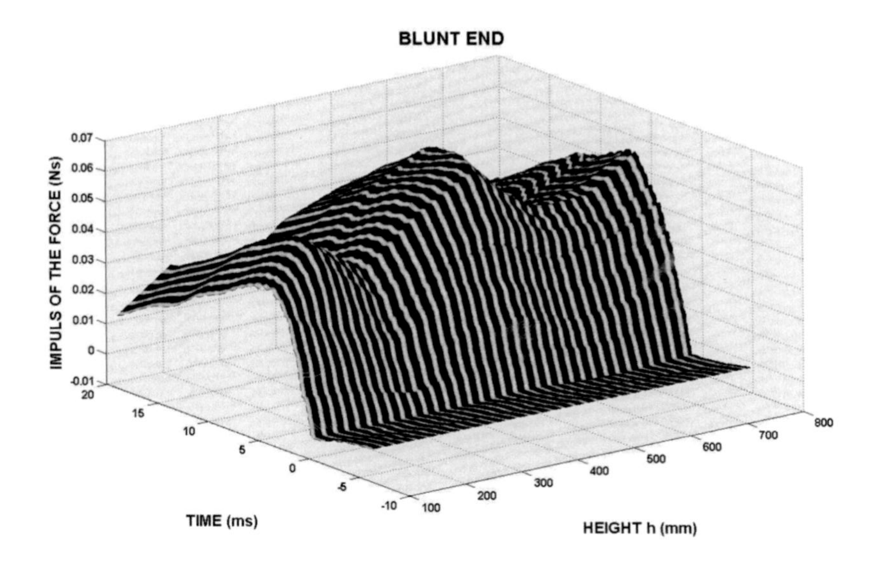

Figure 57. Influence of the fall height on the time history of the impulse of the force.

The next information on the force can be obtained in the Frequency domain. The analysis in the frequency domain is based on the Fourier transform (see e.g. Stein et al., 2003)

For a continuous function of one variable *f(t)*, the Fourier Transform *F(ω)* is defined as:

$$F(\omega)=\int_{-\infty}^{+\infty} f(t)e^{-i\omega t}\,dt \tag{18}$$

And the inverse transform as:

$$f(t)=\int_{-\infty}^{+\infty} F(\omega)e^{i\omega t}\,d\omega\,, \tag{19}$$

where F is the spectral function and ω is the angular frequency.

The same procedure can be used for the Fourier transform of a series *x(k)* with *N* samples. This procedure is termed as the discrete Fourier Transform (DFT). A special kind of this transform is Fast Fourier Transform

(FFT). This procedure is a part of the most of software packages dealing with the signal processing.

This algorithm is a part of the MATLAB software. An example of the spectral function amplitude is shown in the Figure 58. This function is characterized by a peak value at relatively low frequency. For the whole spectrum, the momentum M_o (Eq. (20), the momentum M_1 (Eq. (21) the central frequency *CF* (Eq. (22) and the variance *Var* (Eq. (23) were calculated (Oppenhein and Schafer, 1989).

$$M_0 = \sum F(\omega)\Delta\omega \tag{20}$$

$$M_1 = \sum F(\omega)\omega\Delta\omega \tag{21}$$

$$CF = \frac{M_1}{M_0} \tag{22}$$

$$Var = \frac{\sum(\omega - CF)F(\omega)}{\sum F(\omega)} \tag{23}$$

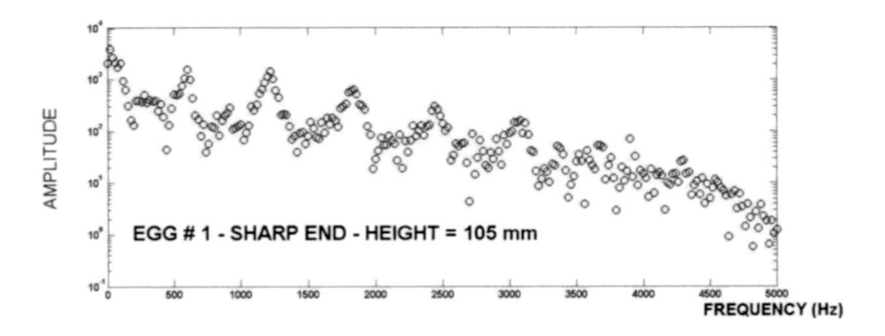

Figure 58. Fourier analysis (FFT) of the force.

The parameters M_o, M_1 and *CF* are given in the Table 15 together with the frequency at which the maximum on the amplitude – force dependence occurs.

Table 15. Characteristics of the amplitude – frequency spectrum – raw eggs

	Egg No	ω (s^{-1})	M_o	M_1	$CF=M_1/M_o$ (s^{-1})	h (mm)
Sharp end	1	20	2230137	1.11E+11	49980	105
	2	20	2447438	1.36E+11	49980	200
	3	20	2717143	1.36E+11	49980	300
	4	20	2611760	1.31E+11	49980	400
	6	20	3067572	1.53E+11	49980	500
	7	20	2800472	1.4E+11	49980	600
	9	20	3325453	1.66E+11	49980	700
	10	20	3106422	1.55E+11	49980	780
Blunt end	11	20	3489978	1.74E+11	49980	780
	12	20	3556307	1.78E+11	49980	780
	13	20	2826837	1.41E+11	49980	700
	14	20	2722207	1.36E+11	49980	600
	15	20	3124508	1.56E+11	49980	500
	16	20	2591775	1.3E+11	49980	400
	17	20	2484916	1.24E+11	49980	300
	18	20	2496437	1.25E+11	49980	200
	19	20	2737091	1.37E+11	49980	105

It is obvious that frequency at which the maximum in contact force frequency spectrum occurs is independent on impact velocity as well as egg position (direction of loading). The same rule is valid for central frequency.

4.2.2. Boiled Eggs

Examples of force-time dependencies in the egg-rod contact point are shown in Figs 59-60. Different fall heights and/or impact velocities are considered. Similar dependencies were obtained for other impact velocities. Concerning qualitative course of the dependencies, the courses are simile as those received for raw eggs.

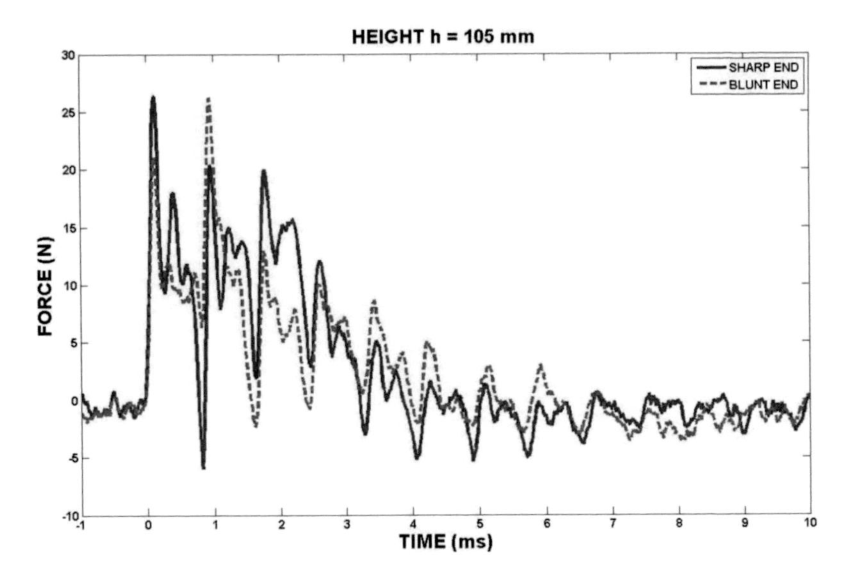

Figure 59. Force-time dependence in the egg-rod contact point.

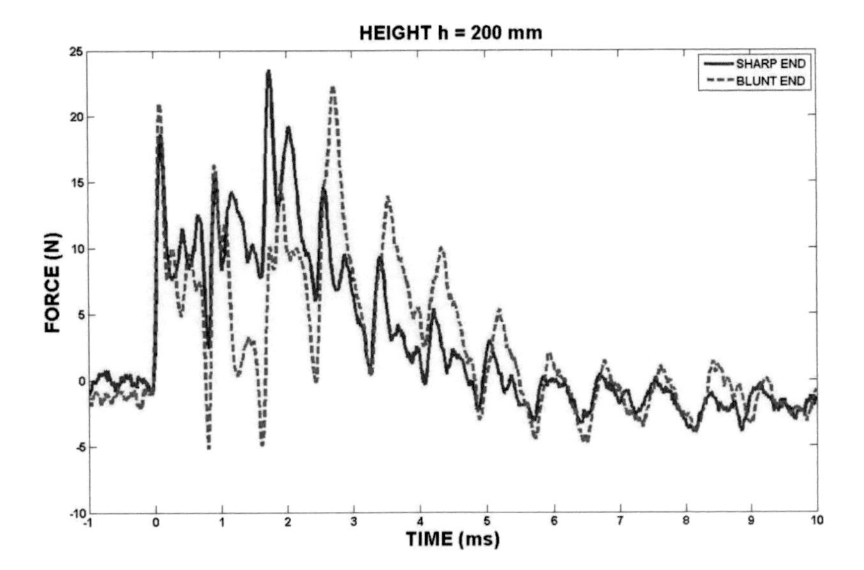

Figure 60. Force-time dependence in the egg-rod contact point.

The force courses in the egg-rod contact point are qualitatively similar as ones recorded for raw eggs – see examples given in Figs 61-62.

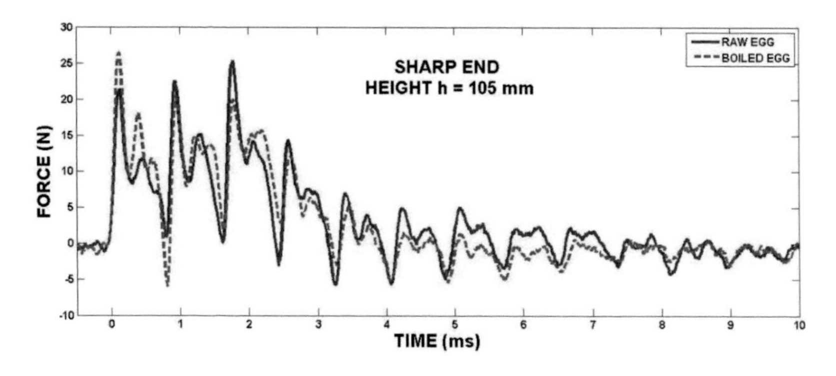

Figure 61. Force-time dependence in the egg-rod contact point.

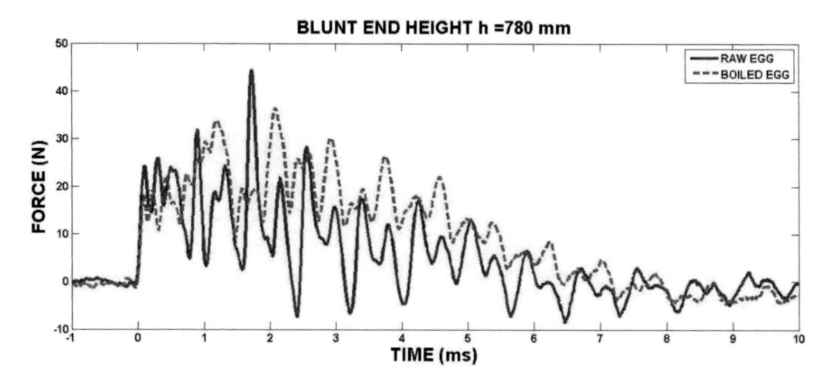

Figure 62. Force-time dependence in the egg-rod contact point.

Following pictures 63 and 64 show the values of contact force impulse. Neither these values indicate considerable influence of thermal treatment on their strain behavior of the eggs.

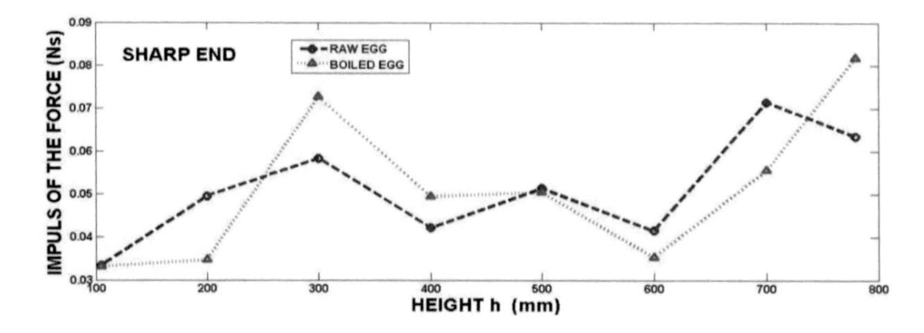

Figure 63. Contact force impuls.

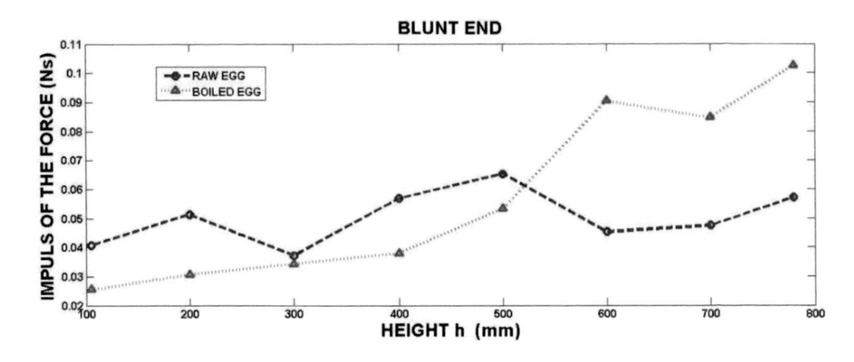

Figure 64. Contact force impuls.

Similar results arise from data obtained in frequency spectrum – see Table 16.

Table 16. Characteristics of the amplitude – frequency spectrum – boiled eggs

	h (mm)	ω_{max} (s-1)	M_0	M_1	CF (Hz)
Sharp end	105	20	2462049	1.23E+11	49980
	200	20	2025868	1.01E+11	49980
	300	20	2721826	1.36E+11	49980
	400	20	2714301	1.36E+11	49980
	500	20	2395353	1.2E+11	49980
	600	20	2382926	1.19E+11	49980
	700	20	3395568	1.7E+11	49980
	780	20	3256423	1.63E+11	49980
Blunt end	105	20	2135069	1.07E+11	49980
	200	20	2303722	1.15E+11	49980
	300	20	2696738	1.35E+11	49980
	400	20	2557428	1.28E+11	49980
	500	20	3283512	1.64E+11	49980
	600	20	3723301	1.86E+11	49980
	600	20	3608312	1.8E+11	49980
	700	20	3843944	1.92E+11	49980
	780	20	3309206	1.65E+11	49980

4.2.3. Peeled Eggs

The eggs were peeled and dropped from different heights in such way that they impacted either on their sharp or blunt end. These impact experiments were performed also for extracted yolk itself. Example record of the force course in the peeled egg-rod contact point is shown in Figure 65. It can be seen that force course includes less oscillations than in the case of egg with shell. Also the oscillation amplitudes are considerably lower. The same results are valid for force courses of falling egg yolk – see Figure 66.

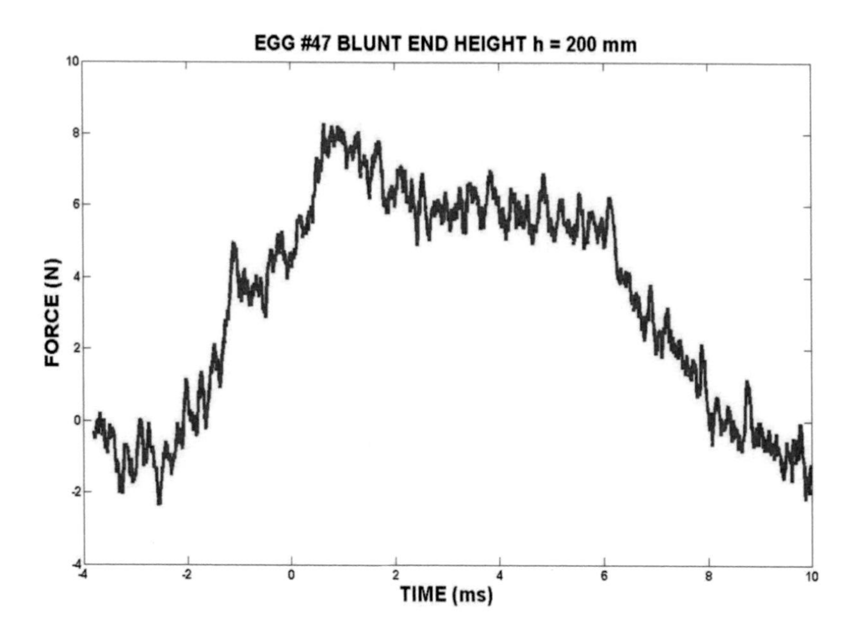

Figure 65. Contact force acting after fall of peeled egg.

Figure 67 shows a comparison of different force records for different fall combinations. A relevant qualitative as well as quantitative difference can be seen for strain behavior of the egg with and/or without eggshell, raw egg and egg yolk. It is obvious that observed oscillations are connected with wave propagation in the eggshell.

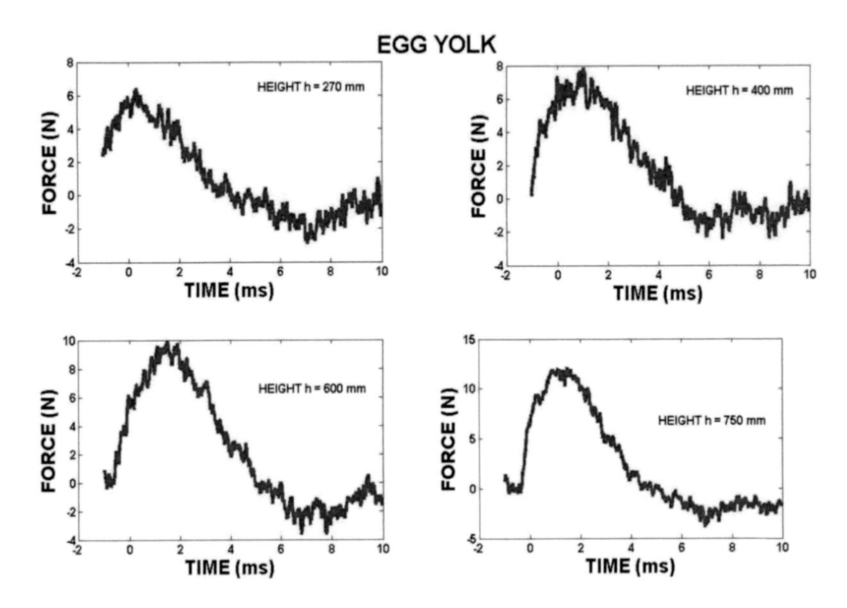

Figure 66. Contact force acting after fall of egg yolk.

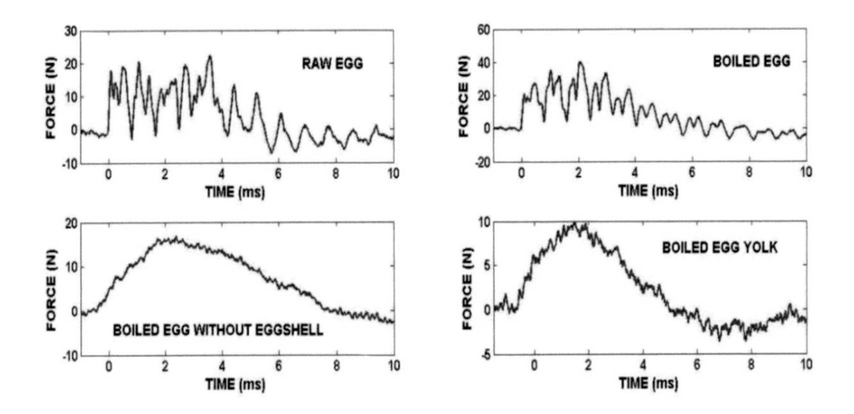

Figure 67. Contact forces acting after fall of raw, boiled, peeled egg, and egg yolk. Height h = 600 mm. Blunt end.

For further description of the problem, a frequency analysis using FFT was performed. Tables 17-19 contain results and basic characteristics obtained from amplitude spectrum of the contact force.

Table 17. Characteristics of the amplitude – frequency spectrum – boiled eggs without eggshell – blunt end

Egg No	h (mm)	ωmax (-1)	M_0	M_1	CF (Hz)
47	200	20	1046892	5.23E+10	49980
48	300	20	1053188	5.26E+10	49980
51	400	20	1394342	6.97E+10	49980
59	600	20	1466786	7.33E+10	49980

Table 18. Characteristics of the amplitude – frequency spectrum – boiled eggs without eggshell – sharp end

Egg No	h (mm)	ωmax (-1)	M_0	M_1	CF(Hz)
60	105	20	958730.5	4.79E+10	49980
57	300	20	1199991	6E+10	49980
56	600	20	1253331	6.26E+10	49980

Table 19. Characteristics of the amplitude – frequency spectrum – boiled yolk

Egg No	h (mm)	ωmax (-1)	M_0	M_1	CF (Hz)
-	270	20	936597	4.68E+10	49980
-	400	20	1033762	5.17E+10	49980
-	600	20	1126646	5.63E+10	49980
-	750	20	1278810	6.39E+10	49980

It is evident that there is no change in frequency, which is the amplitude maximum reached in. These spectra are shown in Figs. 68-70.

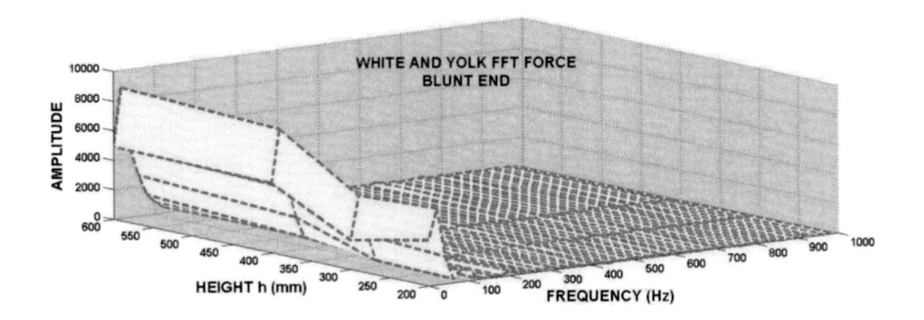

Figure 68. Amplitude frequency spectrum of the contact force.

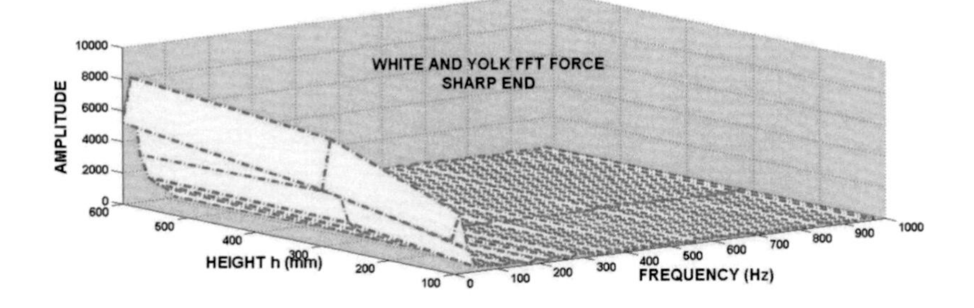

Figure 69. Amplitude frequency spectrum of the contact force.

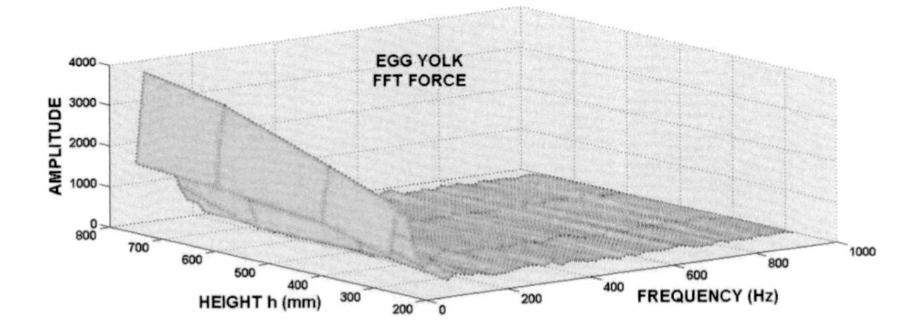

Figure 70. Amplitude frequency spectrum of the contact force.

The differences are largely notable especially in case of M_0 and M_1 – see Figs. 71-72.

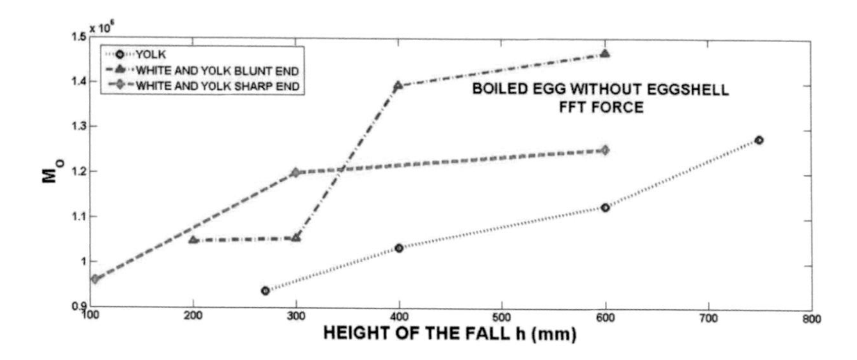

Figure 71. Size of moment M_0.

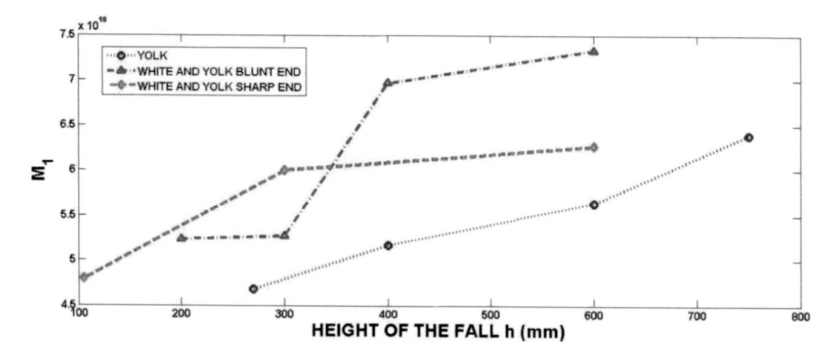

Figure 72. Size of moment M_1.

More detailed analyses of the interaction between egg and rod (or stop block generally) is conditioned by gaining more data, e.g. by use of high-speed camera. Such data would complete the information on impact behavior of eggs and could serve as an input for evaluation of numerical methods reliability and creating of numerical simulations of given problems. Such approaches will be the subject of future research.

Chapter 5

CONCLUSION

Presented chapter contains overview of the basic results received within the research focused on impact loading of the hen's eggs. Following problems have been solved:

- Non-destructive ball impact on the eggshell resulting in data collecting for evaluation of non-destructing eggshell behavior.
- Loading of the eggs by instrumented bar resulting in determination of the eggshell strength.
- Fall of the egg on the rigid stop, which was represented by rod enabling recording of the contact forces. This research was aimed on collecting the data for reliable evaluation of given problem numerical simulation applicability. It the applicability of numerical simulation will be confirmed, generalization of the egg impacting would be possible and different conditions such as transport or storing manipulation could be simulated.

Following results have been obtained for above listed research problems:

Ad 1. The detail analysis of the eggshell response to the non-destructive impact of the ball led to obtaining of large amount of new knowledge concerning egg acoustic behavior. The egg dynamic resonance frequency was detected. It was obtained through the analysis of the dynamically measured frequency response of an excited egg. The effect of excitation point, detected point, impact intensity, excitation material, different egg mass, density, shell stiffness, and shell crack on the dominant frequency were assessed. The basic results are summarized below:

- The dominant frequency was significantly affected by the shell crack, shell stiffness, egg mass and egg density.
- The excitation point, detected point, excitation velocity, and the impacting material did not significantly affect the dominant frequency.
- The dominant frequency increased with increase of the shell stiffness and/or the density and decreased with increase of the egg mass.

The numerical model of the egg has been also proposed. Although this model uses many simplified assumptions, the good agreement between experimental and numerical results suggests that there is a good chance to obtain reliable model. The proposed numerical model can minimize the number of experiments but it can also serve as a tool for describing of many other dynamic loading conditions.

Ad 2. The experimental method of the free falling rod has been developed for the measurement of the eggshell strength under dynamic loading conditions. This method enables simultaneous evaluation of the eggshell strength and acoustic response of the tested egg. The effort in this study was limited on determination of the dynamic strength value. It has been found that the force at which the eggshell rupture started was significantly dependent on the position of the rod impact. Contrary to the results of the static loading tests, there is a significant difference between the rupture force at the sharp and blunt end. This force is also higher than that obtained at the static loading. This phenomenon, which is common for most of metallic materials, polymers and also some brittle materials, must be verified and explained in terms of the eggshell microstructure.

Because the values of the eggshell strength expressed in terms of the rupture force are affected by many factors (egg specific gravity, egg mass, egg volume, egg surface area, egg thickness, shell weight, shape and shell percentage) it is necessary to perform the numerical simulation of these experiments in order to avoid the simultaneous influence of these factors. The numerical model of the given experiment has been proposed. The model respects the true shape of the eggshell. Preliminary results of the numerical simulation show reasonable agreement with experimental data. The tensile stresses which correspond to the impact conditions under which the eggshell damage starts have been determined. It seems that this quantity may represent an intrinsic property of the eggshell. Verification of this hypothesis needs both, further experiments and numerical computations and some other experiments with rod of different diameters and rod tip shapes (conic, ball etc.). Instead of further research focused on numerical analysis, some other problems should be solved in order to improve capabilities of the experimental method described in this chapter. The main problem consists in the use of microscopic method for the eggshell defects observation.

Ad 3. In this part of a research, experimental method enabling detection of time history of contact force between an egg and rigid stop was implemented. The rigid stop was represented by the elastic rod of a circular cross-section equipped with the strain gauges used for force detecting. The advantage of this approach consists in possibility of testing of different rigid stop materials. In this case, the rod was made of PMMA material. Both, the raw and boiled eggs were examined within this research. Following results have made:

- Time course of the contact force is connected with numerous oscillations of large amplitude which are comparable with force maximum. Qualitative character of this value's time history is analogous for both, raw and boiled eggs fall on the sharp as well as blunt end and it is also independent on impact velocity. Above mentioned variation reflects the size of the moment, which is the area under force-time curve.
- The frequency spectrum of the force pulse is characteristic with its marked and noticeable maximum. The maximum is reached in the frequency which is independent on impact velocity, egg orientation during the impact (sharp or blunt end). The frequency is the same for both, raw and boiled eggs.
- Fall of the peeled egg is connected with much smaller oscillations of egg-rod contact forces and relatively low oscillation amplitudes. The value of contact force is obviously lower than in the case of unpeeled egg. The same result is valid for the case of yolk fall. It is evident that presence of the oscillations is connected with eggshell. The oscillations are probably not developed as a consequence of egg fluids movement, because qualitatively similar results were obtained for both, raw and boiled eggs. To a certain extent it is also possible to assume that development of these oscillations is not connected with crack creation and/or gradual eggshell breaking.

This result is based on the fact that oscillation character is independent on impact velocity and thus damage extent. It is evidently the consequence of wave propagation and interference in the material.

Following result can be pronounced as a general rule: Vibrational characteristics during non-destructive egg impact loading are connected with egg geometry and dimensions, egg mass, eggshell thickness and strength properties. Determination of these characteristics is considerably significant for both, practical applications (e.g. automatic egg sorting) and basic research. Aforementioned values are dependent e.g. on eggshell structure and can be used as a data for evaluating of feed quality effect and/or cracks presence evaluation.

Importance of the fall experiments consists especially in the possibility of reliability verification of the egg models in the different numerical softwares. Development of the numerical simulation is essential for evaluation of wide range of problems connected with egg mechanical loading (egg collecting, transport, storing etc.). It is obvious that sole experiments cannot cover and implicate all possible variations of aforementioned influences – for practical, time consuming and economical reasons. One of the main goals of the work presented in this chapter was to point out and refer to potential perspectives in this problematic.

ACKNOWLEDGMENT

The research was supported by the Grant Agency of the Czech Academy of Science under contract No IAA 201990701.

REFERENCES

Altuntas, E. & Sekeroglu, A. (2008). Effect of shape index on mechanical properties of chicken eggs. *J. Food Eng., 85,* 606-612.

Ar, A. & Rahn, H. (1980). Water in the avian egg: overall budget of incubation. *American Zoologist, 20,* 373-384.

Bain, M. M. (1990). Eggshell strength - a mechanical/ ultra structural evaluation. PhD Thesis, University of Glasgow.

Baryeh, E. A. & Mangope, B. K. (2003). Some physical properties of QP- 38 variety pigeon pea. *Journal of Food Engineering, 56,* 59-65.

Bell., D. (1984). Egg breakage - From the hen to the consumer. *Calif. Poult. Lett.,* 2-6.

Bliss, G. N. (1973). *Crack detector*, U.S. Patent 3744299.

Brooks, J. & Hale, H. P. (1955). Strength of the shell of the hen's egg. *Nature*, 175, 848-849.

Buchar, J. & Simeonovova, J. (2001). On the identification of the eggshell elastic properties under quasistatic compression. In *19th CAD – FEM USERS MEETING 2001-International Congress on FEM Technology.* 17-19. October, Potsdam, Berlin Germany, Vol. 2, pp 1-8. Published by CAD – Fem GmbH, Munchen.

Carter, T. C. (1970). The hen's egg: factors affecting the shearing strength of shell material. *Br. Poult. Sci., 11,* 433-449.

Carter, T. C. (1971). The hen's egg: variation in tensile strength of shell material and its relationship with shearing strength. *Br. Poult. Sci., 12,* 57-76.

Carter, T. C. (1976). The hen's egg: Shell forces at impal and quasi–static compression, *Br. Poult. Sci., 17,* 199-214.

Coucke, P. (1998). Assessment of some physical egg quality parameters based on vibration analysis. PhD Thesis. Katholieke Universiteit Leuven. Belgium.

Coucke, P., Dewil, E., Decuypere, E., & De Baerdemaeker, J. (1999). Measuring the mechanical stiffness of an eggshell using resonant frequency analysis. *Br. Poult. Sci., 40*, 227-232.

Chung, R. A. & Stadelman, W. J. (1965). A study of variations in the structure of the hens egg. *British Poultry Science, 6,* 277-282.

De Ketelaere, B., Coucke, P., & De Baerdemaeker, J. (2000). Eggshell crack detection based on acoustic resonance frequency analysis. *J. Agric. Eng. Res., 76,* 157-163.

De Ketelaere, B., Govaerts, T., Coucke, P., Dewil, E., Visscher, J., Decuypere, E., & De Baerdemaeker, J. (2002). Measuring the eggshell strength of 6 different genetic strains of laying hens: techniques and comparisons. *British Poultry Science, 43,* 238-244.

De Ketelaere, B., Vanhoutte, H., & De Baerdemaeker, J. (2003). Parameter estimation and multivariable model building for the non-destructive on-line determination of the eggshell strength. *Journal of Sound and Vibration, 266,* 699-709.

Denys, S., Pieters, J. G., & Dewettinck, K. (2003). Combined CFD and experimental approach for determination of the surface heat transfer coefficient during thermal processing of eggs. *Journal of Food Science, 68,* 943-951.

Dunn, I. C., Bain, M., Edmond, A., Wilson, P. W., Joseph, N., Solomon, S., De Ketelaere, B., De Baerdemaeker, J., Schmutz, M., Preisinger, R., & Waddington, D. (2005a). Heritability and genetic correlation of measurements derived from acoustic resonance frequency analysis; a novel method of determining eggshell quality in domestic hens. *British Poultry Science, 46,* 280-286.

Dunn, I. C., Bain, M., Edmond, A., Wilson, P. W., Joseph, N., Solomon, S., De Ketelaere, B., De Baerdemaeker, J., Schmutz, M., Preisinger, R., & Waddington, D. (2005b). Dynamic stiffness (K_{dyn}) as a predictor of eggshell damage and its potential for genetic selection. In *Proc. XIth European Symposium on the Quality of Eggs and Egg Products*. Doorwerth. The Netherlands. 23-26 May 2005.

Erdogdu, F., Sarkar, A., & Singh, R. P. (2005). Mathematical modeling of air-impingement cooling of finite slab shaped objects and effect of spatial variation of heat transfer coefficient. *Journal of Food Engineering, 71,* 287-294.

Erdogdu, F., Ferrua, M., Singh, S. K., & Singh, R. P. (2007). Air-impingement cooling of boiled eggs: Analysis of flow visualization and heat transfer. *Journal of Food Engineering, 79,* 920-928.

Hammmerle, J. R. & Mohsenin, N. N. (1967). Determination and analysis of failure stresses in egg shells. *J. agric. Engng Res., 12,* 13-21.

Hunton, P. (1993). Understanding the architecture of the eggshell. In *Proc. 5th Eur. Symp. Qual. Eggs and Egg Prod*, pp 141–147, Tours, France. World's Poult. Sci. Assoc., Cedex, France.

Ju, B. B., Liu, K. K., Ling, S. F., & NG, W. H. (2002). A novel technique for characterizing elastic properties of thin biological membrane. *Mechanics of Materials, 34,* 749-754.

MacLeod, N., Bain, M. M., & Hancock, J. W. (2006). The mechanics and mechanisms of failure of hens' eggs. *Int. J. Fracture, 142,* 29-41.

Moayeri, A. (1997). Probe, device and method for testing eggs. U.S. Patent 5728939.

Mohsenin, N. N. (1970). *Physical properties of plant and animal materials.* New York: Gordon and Breach Science Publishers.

Narushin, V. G. (2001). Shape geometry of the avian egg. *Journal of Agricultural Engineering Research, 79,* 441-448.

NEDOMOVÁ, Š., SEVERA, L., & BUCHAR, J. (2009). Influence of hen egg shape on eggshell compressive strength. *Int. Agrophysics, 23,* 249-256.

Olsson, N. (1934). *Studies on Specific Gravity of Hen's Egg.* A Method for Determining the Percentage of Shell on Hen's Eggs. Otto Harrassowitz. Leipzig. Germany.

Oppenhein, A. V. & Schafer, R. W. (1989). *Discrete-time signal processing.* New Jersey: Prentice Hall International, Inc.

Reissner, E. (1946). Stresses and small displacements of shallow spherical shells. *Int. J. Math. Phys., 25,* 80-85.

Rodriguez-Navarro, A., Kalin, O., Nys, Y., & Garcia-Ruiz, J. M. (2002). Influence of the microstructure and crystallographic texture on the fracture strength of hen's eggshells. *Br. Poult. Sci, 43,* 395-403.

Sinha, D. N., Johnston, R. G., Grace, W. K., & Lemanski, C. L. (1992). Acoustic resonances in chicken eggs. *Biotechnology Progress, 8,* 240-243.

Sluka, S. J., Besch, E. L., & Smith, A. H. (1967). Stresses in impacted egg shells. *Trans. Am. Soc. Agric. Engrs, 10,* 364-369.

Sugino, H., Nitoda, T., & Juneja, L. R. (1997). General chemical composition of hen eggs. In Yammamoti, T., Juneja, L. R., Hatta, H.

and Kim, M. (Eds). *Hen eggs.* Their basic and applied science. CRC Press.

Thompson, B. K., Hamilton, R. M. G., & Voisey, P. W. (1981). Relationships among various traits relating to shell strength, among and within five avian species. *Poult. Sci., 60,* 2388-2394.

Timoshenko, S. P. & Woinowsky-Krieger, S. (1959). *Theory of Plates and Shells.* New York, McGraw-Hill Book Company, Inc.

Tung, M. A., Staley, L. M., & Richards, J. F. (1969). Estimation of Young's modulus and failure stresses in the hen's egg shell. *Can. Agric. Engng, 11,* 3-5.

Tyler, C. & Geake, F. H. (1963). A study of various impact and crushing methods used for measuring shell strength. *Br. Poult. Sci., 4,* 49-61.

Voisey, P. W. & Hamilton, R. M. G. (1976). Factors affecting the non-destructive and destructive methods of measuring eggshell strength by the quasi-static compression test. *Br. Poult. Sci,. 17,* 103-124.

Voisey, P. W. & Hunt, J. R. (1967a). Physical properties of egg shells. Stress distribution in the shell. *Br. Poult. Sci. 8,* 263-271.

Voisey, P. W. & Hunt, J. R. (1967b). Relationship between applied force, deformation of eggshells and fracture force. *J. Agric. Eng. Res., 12,* 1-4.

Voisey, P. W. & Hunt, J. R. (1969). Effect of compression speed on the behaviour of eggshells. *J. agric. Engng Res., 14,* 40-46.

Voisey, P. W. & Hunt, J. R. (1974). Measurement of eggshell strength. *J. Texture Stud., 5,* 135-182.

Wells, R.G. (1968). Egg quality Characteristics. In T.C. Carter, ed., *Egg Quality: A Study of the Hens Eg.* (pp. 214- 225), Edinburgh: Oliver and Boyd.

Yang, S. H., Hsi-Lung, C., & CHUNGHWA, W. (1995). Quality control on thousand year egg by vibration identification, In *Proceedings of the 13th International Modal Analysis Conference*, Nashville, TN, pp. 1242-1247.

INDEX

A

B

C

D

E

F

G

H

I

L

M

N

O

P

R

S

T

V

W

X

Y